The George Washington University
Astronomy 1002
SCALE-UP Workbook

Author: Bethany Cobb Kung
3rd Edition: Fall, 2019

Table of Contents

Chapter 1: A Modern View of the Universe	3
Chapter 3: The Science of Astronomy	9
Chapter 4: Making Sense of the Universe	15
Chapter 5: Light and Matter	27
Chapter S4: Building Blocks of the Universe	35
Chapter 14: Our Star	45
Chapter 15: Surveying the Stars	51
Chapter 16: Star Birth	61
Chapter 17: Star Stuff	67
Chapter S2: Space and Time	75
Chapter S3: Spacetime and Gravity	85
Chapter 18: The Bizarre Stellar Graveyard	93
Chapter 19: Our Galaxy	101
Chapter 20: Galaxies and the Foundation of Modern Cosmology	109
Chapter 22: The Birth of the Universe	117
Chapter 23: Dark Matter, Dark Energy, and the Fate of the Universe	123

Astronomy 1002 SCALE-UP Workbook, Third Edition
Copyright © 2019 by The Department of Physics, The George Washington University

All rights reserved. No part of this publication may be reproduced or transmitted in any form or by any means, electronic or mechanical, including photocopying, recording, or any information storage and retrieval system, without the written permission of the publisher.

Requests for permission to make copies of any part of the work should be mailed to:

Permissions Department
Academx Publishing Services, Inc.
P.O. Box 208
Sagamore Beach, MA 02562
http://www.academx.com

Printed in the United States of America

ISBN-13: 978-1-68284-626-1
ISBN-10: 1-68284-626-1

Chapter 1

1.1) Obtain a pack of "astronomical object cards" from the instructor.
   a) Arrange the astronomical objects on each card according to **size**. Record this order below.

   b) Arrange the astronomical objects on each card according to **hierarchy**. Record this order below.

1.2)

Object	Diameter
Sun	1.39×10^6 km
Milky Way Galaxy	1×10^5 ly

Objects	Separation
Sun to Alpha Centauri	4×10^{13} km
Milky Way to Andromeda	2.5×10^6 ly

a) If you use marbles (approximately 1 cm in diameter) to represent **stars**, how far apart would you need to place these marbles to represent **to scale** the distance between the Sun and Alpha Centauri (a nearby star)? Start this problem by completing the ratio set-up below – use "x" to indicate the unknown quantity. Be sure to include *__units__* every time you write down a value.

$$\frac{\text{(actual diameter)}}{\text{(scaled diameter)}} = \frac{\text{(actual separation)}}{\text{(scaled separation)}}$$

Scaled Separation = _____ cm

Scaled Separation = _____ km

b) If you use marbles (approximately 1 cm in diameter) to represent **galaxies**, how far apart would you need to place these marbles to represent **to scale** the distance between the Milky Way and the Andromeda Galaxy?

Scaled Separation = _____ cm

c) How does the separation between stars within a galaxy (relative to the size of a typical star) compare to the separation between individual galaxies (relative to the size of a typical galaxy)?

d) Which is more likely to **collide** in space – two stars within a galaxy, or two individual galaxies? *Explain your choice.*

1.3) The circle below is approximately 10 cm in diameter. Within this circle, you will draw a scale model of our Galaxy (from the viewpoint of looking down on the Milky Way from far above). The true diameter of the Milky Way is approximately 1×10^5 light-years.
   a) The central bulge of the Milky Way is about 2×10^4 light-years in diameter. Calculate the approximate scaled size of the central bulge, and sketch the bulge below **to scale**.

   b) Add some spiral arms to the sketch.
   c) Our solar system is located around 2.7×10^4 light-years from the center of the Milky Way. Calculate and mark the position of the solar system in the sketch **to scale**.

d) Our solar system out to Pluto is about 100 AU in diameter. Using a standard pencil could you correctly mark **to scale** the size of the solar system on your Galaxy model on the previous page? YES or NO *(circle one)*. Explain your answer. When you are asked to explain an answer in this workbook, be sure to *always* include evidence or data to support your answer. In this case, the strongest evidence would be a calculation of the scaled-down size of the solar system and a statement about whether or not it would be reasonable to draw a mark of that size with a pencil.

e) The solar system completes one orbit of the Milky Way every 230 million years. What is the orbital speed of the solar system around the galaxy in units of km/hr? When you are asked to complete a calculation in this workbook, be sure to *show your work* (though, of course, you may determine final numerical values using your calculator), *label all units*, and *report final values to a reasonable number of significant digits*.

1.4) The image below represents 8 imaginary solar systems. You are an observer in the central solar system ("O"). Imagine that people (or aliens) in the other solar systems are all attempting to contact you using extremely powerful laser beam signaling devices.

a) If the signals were sent <u>simultaneously</u> toward you, in what order would you receive them?

First __E__ __C__ __D__ __A__ __F__ __B__ __G__ *Last*

b) Of course, the signals are unlikely to be so well coordinated. Imagine instead that the signals from A, B and G are sent first (in the year corresponding to Earth-year 2010), then 15 years later (Earth-year 2025) the signal from both C and F are sent, then 8 years later (Earth-year 2033) the signal from D is sent, then 2 years later (Earth-year 2035) the signal from E is sent. In what order will you <u>receive</u> these signals, and in which Earth years?

First __A__ in year __2030__

__C__ in year __2036__

__B__ in year __2040__

__E__ in year __2045__

__D__ in year __2048__

__F__ in year __2050__

Last __G__ in year __2055__

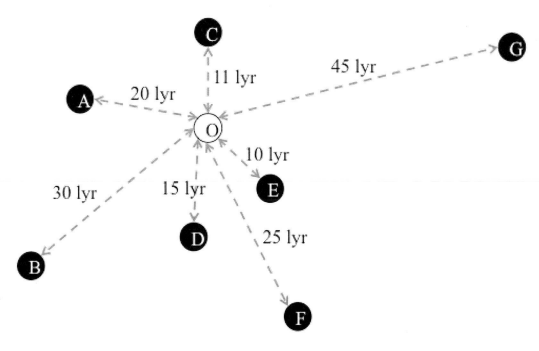

Chapter 3

3.1) What are the three hallmarks of science? How does the pseudoscience of astrology **fail** to fulfill each hallmark? Imagine any one revolutionary discovery/theory in science that you are familiar with (examples: Copernican revolution, germ theory, plate tectonics, quantum mechanics, evolution, global climate change, etc.). How does that particular revolution **succeed** in fulfilling each of the three hallmarks?

Hallmark #1:

    Astrology fails because…

    But _____ succeeds because…

Hallmark #2:

    Astrology fails because…

    But _____ succeeds because…

Hallmark #3:

    Astrology fails because…

    But _____ succeeds because…

3.2) Which of the following statements use the words hypothesis or theory in a <u>scientific</u> manner? If the word is being used more colloquially, what does the speaker really mean? Suggest an alternative word or phrase for them to use.

Statement	Scientific? (*circle one*)	Alternative word or phrase
"I **hypothesize** that blue light has a shorter wavelength than red light."	YES NO	
"Your **hypothesis** is that bacteria can cause diseases such as meningitis."	YES NO	
"Janet **hypothesizes** that the movie she is about to see is going to be awesome."	YES NO	
"The **theory** of quantum mechanics makes predictions that can be tested."	YES NO	
"My **theory** is that Jane didn't come to class today because she still has to write a 7-page paper for her history class."	YES NO	
"Archeologists **theorize** that the people native to this land died out because of diseases brought by foreign traders."	YES NO	

3.3) Bob says to you: "Cats are smarter than dogs." Could this claim be evaluated scientifically or does it fall into the realm of non-science? *Explain your answer.*

3.4) **Think like a scientist.** You are part of a scientific team investigating stars in the Milky Way.
  a) A colleague provides you with the following histogram, which displays the mass of all the stars in the 100 star systems that are nearest to the Earth. Note that many star systems contain more than one star, so the mass of 143 different stars is plotted. Your colleague says, "From this evidence, I conclude that stars of greater than 2 solar masses do not exist in the Milky Way Galaxy." Does the evidence given below support that conclusion? *Explain your reasoning*.

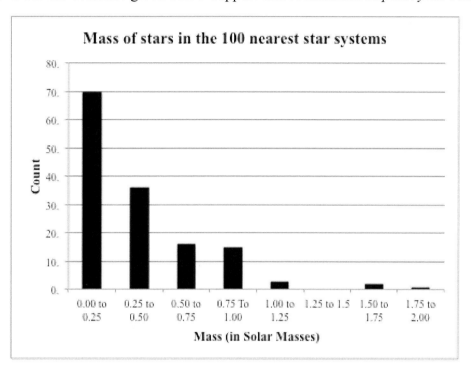

b) Consider the data below. What conclusions can you draw from the evidence provided? *Explain your reasoning*. Draw a graph of the data which supports your conclusions.

Star	Mass (in Solar Masses)	Radius (in Stellar Radii)
Orionis C	18	5.9
Becrux	16	5.7
Spica	10.5	5.1
Betelgeuse	10	1,000
Achernar	5.4	3.7
Rigel	3.5	2.7
Fomalhaut	2.2	2
Altair	1.9	1.8
Polaris A	1.6	1.5
Procyon A	1.35	1.2
The Sun	1	1
Sirius B	0.98	0.01
Epsilon Eridani	0.78	0.79
Lalande 21185	0.33	0.36
Ross 128	0.2	0.21
Wolf 359	0.1	0.12

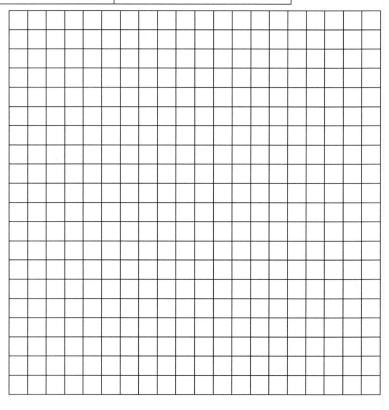

c) You are working on the following research question: "How does a star's temperature relate to its surface color?" Describe in detail what data you would need to collect in order to answer this question – **be specific**! How many data points might you need? How would you choose which stars to consider? *Explain your reasoning*. Also detail exactly how you would analyze this data in order to answer your research question.

Chapter 4

4.1) Speed vs. velocity vs. acceleration.

a) Which phrases below indicate **speed** and which indicate **velocity**? *(Check the appropriate column.)*

Phrase	Speed?	Velocity?
"The fastball was thrown at 40 m/s."		
"The car was heading north at 50 km/hr."		
"The astronauts is moving at 3×10^4 km/s toward Mars."		
"The cheetah's sprint maxed out at 60 mph."		

b) The graph to the right shows the speed of an object over time – *the object is traveling in a straight line, so its direction never changes*. On the graph, circle the letter of each time segment (A – F) during which the object experiences **acceleration**. For each segment in which acceleration is occurring, indicate if the rate of acceleration is **constant** or **changing with time**.

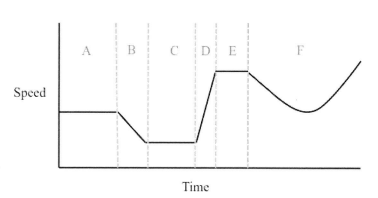

c) In which of the following scenarios is an **acceleration** occurring? *(Circle yes/no.)*

Scenario	Acceleration?
A bicyclist travels along a flat road at 20 m/s.	YES NO
The Earth orbits the Sun.	YES NO
A car slows from 100 m/s to 25 m/s.	YES NO
A chunk of ice falls off a glacier into the ocean below.	YES NO
An elevator travels upward, passing by 1 floor every 3 seconds.	YES NO
A limousine navigates around a traffic circle at 20 mph.	YES NO
A jet plane increases speed by 12 m/s every second.	YES NO
An astronaut in the International Space Station orbits the Earth.	YES NO
An elevator stopped on the 1st floor begins moving toward the 2nd floor.	YES NO
An elevator's cable snaps, and the elevator car begins to free-fall.	YES NO

4.2) Which has more momentum? *(Circle one object in each row.)* In some cases, you will need to perform a calculation (momentum = mass × velocity) to make this determination. Consider units carefully.

A stationary hippopotamus (1500 kg).	OR	A bullet (0.01 kg) fired at 300 m/s.
A 2000 kg car moving at 50 mph.	OR	A 2000 kg car moving at 100 mph.
A 2000 kg car moving at 100 mph.	OR	A 1000 kg car moving at 100 mph.
A 1000 kg car moving at 50 mph.	OR	A 2000 kg car moving at 100 mph.
An elephant (5000 kg) walking at 0.25 m/s.	OR	A bullet (0.01 kg) fired at 400 m/s.
A sprinter (65 kg) running at 11 m/s.	OR	A bicyclist (90 kg) cycling at 9 mph.

4.3) How would your weight feel in each of the following scenarios? Would you feel heavier than normal, lighter than normal, or weightless? *(Circle one in each row.)*

Scenario	Weight?		
While falling off a tall building (don't worry – you're falling toward a pile of soft mattresses!)	Heavier	Lighter	Weightless
In a Space Shuttle, blasting off from the surface of the Earth.	Heavier	Lighter	Weightless
In an elevator that has just started moving downward.	Heavier	Lighter	Weightless
In an elevator that has just started moving upward.	Heavier	Lighter	Weightless
Standing on the surface of the Moon.	Heavier	Lighter	Weightless
Standing on the surface of Mars.	Heavier	Lighter	Weightless
Standing on the surface of a planet that is about the same size as the Earth, but 25% more massive.	Heavier	Lighter	Weightless
In a space ship that is accelerating toward Mars at 15 m/s^2.	Heavier	Lighter	Weightless
In a space ship that is headed toward Mars at a constant velocity of 100 km/s.	Heavier	Lighter	Weightless
In the International Space Station, orbiting the Earth.	Heavier	Lighter	Weightless
While parachuting – before you've opened your parachute.	Heavier	Lighter	Weightless
At the top of a spinning Ferris wheel.	Heavier	Lighter	Weightless
At the bottom of a spinning Ferris wheel.	Heavier	Lighter	Weightless
In a missile that is accelerating toward the surface of the Earth at a rate of 30 m/s^2 (this might not end well for you…).	Heavier	Lighter	Weightless

4.4) Consider the equation for Newton's universal law of gravitation:

$$F_g = G \frac{M_1 M_2}{d^2}$$

a) Imagine that you have two objects, separated by a given distance. How would the gravitational force between the two objects change if the system were altered in the following ways? Indicate if the gravitational force would increase or decrease and calculate the factor by which it would change. The first row has been completed for you as an example. Note that *increases* will always be associated with numerical factors >1 and *decreases* with numerical factors <1).

Example: You double the mass of one of the objects.

$$F_g(Initial) = G \frac{M_1 M_2}{d^2}$$

$$F_g(Final) = G \frac{(2 \times M_1) M_2}{d^2} = 2 \times G \frac{M_1 M_2}{d^2} = 2 \times F_g(Initial)$$

therefore $F_g(Final) = 2 \times F_g(Initial)$
→ the gravitational force has increased by a factor of 2

Scenario	Gravitational Force Increase/Decrease?	Numerical Factor
You double the mass of one of the objects.	**Increase** Decrease	2
You double the mass of both of the objects.	Increase Decrease	4
You triple the mass of one of the objects.	Increase Decrease	3
You triple the mass of both of the objects.	Increase Decrease	9
You halve the mass of one of the objects.	Increase **Decrease**	½
You halve the mass of both of the objects.	Increase Decrease	¼
You increase the separation between the objects by a factor of three.	Increase Decrease	1/9
You decrease the separation between the objects by a factor of three.	Increase Decrease	9
You halve the separation between the objects and triple the mass of both objects.	Increase Decrease	36
You double the separation between the objects and double the mass of one of the objects.	Increase Decrease	½
You double the separation between the objects and double the mass of both of the objects.	Increase Decrease	1

b) What objects in the classroom are you currently gravitationally attracted to? Are there any objects in the classroom to which you are not currently gravitationally attracted?

c) Why don't you notice the gravitational attraction between yourself and everyday objects like chairs, desks and other human beings? Obviously, the masses involved are small, but so are the distances, so **why** is the force of gravity between you and these objects imperceptible?

d) A common misperception is that "there is no gravity in space." Based on Newton's universal law of gravitation, why must this statement be false? (Consider how far away from the Earth an object would have to be to experience absolutely no gravitational attraction to the Earth.)

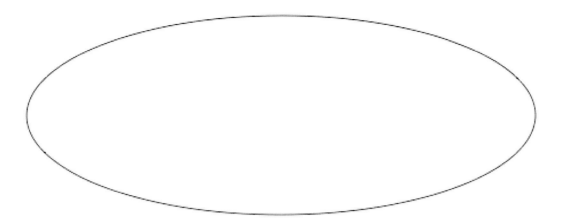

4.5) Consider the orbit above.
   a) In the orbit, draw and label the **major axis**, the **minor axis**, and the **semi-major axis**.
   b) Eccentricity describes how much an ellipse deviates from a perfect circle. Is the shape of the above orbit best described as *round, moderately eccentric,* or *highly eccentric*?

4.6) Consider the equation for Newton's Version of Kepler's Third Law. What does each variable in this equation stand for and what units **must** be associated with each value?

$$p^2 = \frac{4\pi^2}{G(M_1 + M_2)} a^3$$

p is _____ in units of _____

a is _____ in units of _____

M_1 is _____ in units of _____

M_2 is _____ in units of _____

G is _____ in units of _____

4.7) Consider the equation for Newton's Version of Kepler's Third Law:

$$p^2 = \frac{4\pi^2}{G(M_1+M_2)} a^3$$

a) Imagine that you have two objects in orbit around one another. How would the system change if the orbit were altered in the following ways? **Assume that the masses remain constant in all scenarios.** Indicate if the period or the semi-major axis increases or decreases and calculate the numerical factor of that change. The first row has been completed for you as an example.

Example: You double the semi-major axis.

$$p(Initial)^2 = \frac{4\pi^2}{G(M_1+M_2)} a^3$$

$$p(Final)^2 = \frac{4\pi^2}{G(M_1+M_2)} (2 \times a)^3 = (2)^3 \times \frac{4\pi^2}{G(M_1+M_2)} a^3 = 8 \times p(Initial)^2$$

Caution! Don't forget that the period is *squared* in the equation… so you must take the square root of both sides of the equation to solve for the period…

$$p(Final) = \sqrt{8 \times p(Initial)^2}$$

therefore $p(Final) = 2.83 \times p(Initial)$
→ the period has increased by a factor of 2.83

Scenario	Period Increase/Decrease?	Numerical Factor
You double the semi-major axis.	Increase Decrease	2.83
You triple the semi-major axis.	Increase Decrease	
You halve the semi-major axis.	Increase Decrease	
Scenario	Semi-major Axis Increase/Decrease?	Numerical Factor
You double the period.	Increase Decrease	
You triple the period.	Increase Decrease	
You halve the period.	Increase Decrease	

b) Calculate the mass of the Sun using Newton's Version of Kepler's Third Law. The semi-major axis of the Earth-Sun system is 1 AU. The period of the Earth around the Sun is 1 year. Follow the steps below to solve this problem.
   1) Identify which variables are known (given) and which are unknown (to be solved for). Write down each known value *including units*.

   KNOWN VARIABLES: UNKNOWN VARIABLES:

   2) Convert all known values into the units required by the equation. In the case of Newton's Version of Kepler's Third Law, the units that must be used are kilograms, meters and seconds – we must stick to these units because the value of the gravitational constant is G = 6.67×10^{-11} **m³ kg⁻¹ s⁻²**.

   p in seconds = _____

   a in meters = _____

   3) **Algebraically** solve the equation for the <u>unknown value</u> *before* plugging in any numbers. In this case, the unknown value is the total mass of the Earth-Sun system. Show your work below.

   $M_1 + M_2 =$

   4) Finally, use your calculator to solve for the total mass of the Earth-Sun system. Be careful with your order of operations, and don't forget any of the exponents! Note that because the Earth is so much less massive than the Sun ($M_{Sun} \gg M_{Earth}$), the total mass of the Earth-Sun system is approximately equal to the mass of the Sun.

   $M_{Sun} + M_{Earth} \approx M_{Sun} =$ _____

4.8) Escape velocity: $v_{escape} = \sqrt{\dfrac{2GM}{R}}$

a) Calculate the escape velocity from the surface of the Earth. The Earth has a mass of 5.97×10^{24} kg and a diameter of 1.28×10^4 km. Follow the steps below to solve this problem.

1) Identify which variables are known (given) and which are unknown (to be solved for). Write down each known value *including units*.

    KNOWN VARIABLES: UNKNOWN VARIABLES:

2) Convert all known values into the units required by the equation. For escape velocity the units that must be used are kilograms, meters and seconds – we must stick to these units because the value of the gravitational constant is $G = 6.67 \times 10^{-11}$ **m³ kg⁻¹ s⁻²**.

M in kg = _____

R in meters = _____ (Don't confuse *diameter* and *radius!*)

3) **Algebraically** solve the equation for the <u>unknown value</u> *before* plugging in any numbers. In this case, v_{escape} <u>is</u> the unknown value, so this step is already complete.

4) Finally, use your calculator to solve for the escape velocity.

v_{escape} *(from Earth's surface)* = _____

b) Calculate the escape velocity from a distance of 1 AU from the Sun ($M_{Sun} = 1.99 \times 10^{30}$ kg). This would be the velocity required for a spacecraft (that had already been launched from the surface of the Earth into space) to exit the Solar System.

v_{escape} *(from 1 AU from the Sun)* = _____

4.9) Consider the diagram above. The object at point "A" starts at rest and then slides across the *frictionless* surface of the ramp from point "A" to point "E".

   a) Describe how the kinetic energy of this object changes with time.

   b) Describe how the gravitational potential energy of this object changes with time.

   c) At what point does the object have maximum kinetic energy and minimum gravitational potential energy?

   d) At what point does the object have maximum gravitational potential energy and minimum kinetic energy?

   e) Describe how the total energy of this object changes with time.

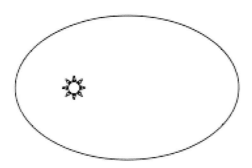

4.10) Consider the orbit above.
   a) Draw and label the position along the planet's orbit where the planet is moving the *fastest*.
   b) Draw and label the position along the planet's orbit where the planet is moving the *slowest*.
   c) Draw and label the position along the planet's orbit where the planet has the *most* kinetic energy.
   d) Draw and label the position along the planet's orbit where the planet has the *least* kinetic energy.
   e) Draw and label the position along the planet's orbit where the planet has the *most* gravitational potential energy.
   f) Draw and label the position along the planet's orbit where the planet has the *least* gravitational potential energy.
   g) How does orbital position affect the total orbital energy of the planet?

4.11) Compare the **temperature** and **thermal energy** in each of the following scenarios. Recall that temperature is a measure of the ***average*** kinetic energy ($E_K = \frac{1}{2} mv^2$) of the particles, while thermal energy is the ***total*** kinetic energy of all the particles. Assume that all the particles mentioned below have the same mass. *(Choose greater than [>], equal to [=], or less than [<] in each column.)*

	Temperature	Thermal Energy	
15 particles, each moving at 500 m/s	> = <	> = <	30 particles, each moving at 500 m/s
100 particles, each moving at 400 m/s	> = <	> = <	100 particles, each moving at 300 m/s
10 particles, each moving at 100 m/s	> = <	> = <	1000 particles, each moving at 10 m/s
1000 particles, each moving at 10 m/s	> = <	> = <	10 particles, each moving at 10,000 m/s
Air molecules in an oven set at 400° F	> = <	> = <	Water molecules in a pot of boiling water.

4.12) Mass-energy equation: $E = mc^2$

a) The atomic bombs dropped on Hiroshima and Nagasaki in 1945 each exploded with as much energy as approximately 15 to 20 kilotons of TNT – this is about 10^8 MegaJoules of energy (for comparison, the typical American household consumes less than 10^5 MJ of energy in an entire *year*). How much mass was converted into energy in each atomic bomb explosion? Follow the steps below to solve this problem.

   1) Identify which variables are known (given) and which are unknown (to be solved for). Write down each known value *including units*.

   KNOWN VARIABLES: UNKNOWN VARIABLES:

   2) Convert all known values into the units required by the equation. For the mass-energy equation the units that must be used are kilograms, meters and seconds – we must stick to these units because we measure energy in the unit of Joules ($J = \frac{kg\, m^2}{s^2}$).

   E in J = _____

   c in m/s = _____

   3) **Algebraically** solve the equation for the <u>unknown value</u> *before* plugging in any numbers. In this case, the unknown value is the mass.

   m =

   4) Finally, use your calculator to solve for the mass.

   m (mass converted into energy) = _____

b) In the Hiroshima atomic bomb, energy was produced when uranium-235 underwent fission. During uranium fission, only about 0.1% of the uranium is converted into energy. How many kilograms of uranium must the Hiroshima bomb have contained?

   total uranium mass in the Hiroshima bomb = _____

(Note the bomb actually contained 64 kg of uranium, but only a small amount of that material actually underwent fission when the bomb was detonated.)

Chapter 5

5.1) Consider the light wave to the right.
  a) Label the wave's **wavelength**.
  b) Draw another light wave with a higher frequency. Label its wavelength.
  c) Draw another light wave with a lower frequency. Label its wavelength.
  d) Which is more energetic light, the wave drawn in 'a', 'b' or 'c'?

5.2) How fast does light travel? Does radio light travel at the same speed as X-ray light? Why or why not?

5.3) Compare the **wavelength, frequency** and **energy** of each type of light. *(Choose greater than [>], equal to [=], or less than [<] in each column.)*

	Wavelength	Frequency	Energy	
Radio waves	> = <	> = <	> = <	Gamma-rays
Red light	> = <	> = <	> = <	Blue light
Ultraviolet light	> = <	> = <	> = <	Infrared light
Optical light	> = <	> = <	> = <	Infrared light
Optical light	> = <	> = <	> = <	Ultraviolet light
X-rays	> = <	> = <	> = <	Optical light
Radio waves	> = <	> = <	> = <	Optical light

5.4) A photon has 4×10^{-19} Joules of energy. What is the **wavelength** of this photon?

5.5) A student has been asked to sketch the approximate blackbody curve for several objects of different temperatures. For each sketch, identify the mistake that the student has made, **then show/draw (on the graph) how the student should modify the graph in order to correct that mistake**.

Object A has a higher temperature than object B.

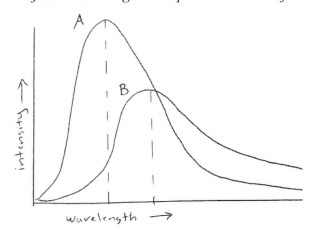

Mistake =

Object C has a higher temperature than object D.

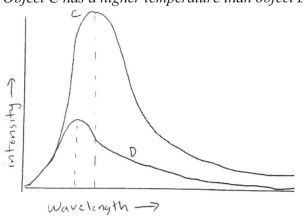

Mistake =

5.6) The typical human being has a body temperature of 37° C.
 a) At what wavelength does a human emit the most thermal radiation? Use Wein's Law to complete this calculation.
$$\lambda_{max} = \frac{2.9 \times 10^6 \text{ nm·K}}{T}$$

 b) Based on the wavelength you calculated above, in what region of the electromagnetic spectrum do humans bodies "glow"?

5.7) Consider the Stefan-Boltzmann law:

$$\text{power per m}^2 = \frac{\text{total power}}{\text{surface area}} = \sigma T^4$$

a) Imagine an object emitting thermal radiation. How would the power emitted per square meter of the object's surface change as the object's temperature changes? Indicate if the change in power per square meter is increased or decreased and calculate the numerical factor of that change. The first row has been completed for you as an example.

Example: You double the temperature.

$P/A(Initial) = \sigma T^4$

$P/A(Final) = \sigma(\mathbf{2} \times \boldsymbol{T})^4 = (2)^4 \times \sigma T^4 = 16 \times P/A(Initial)$

therefore $P/A(Final) = 16 \times P/A(Initial)$

→ the power emitted per area has increased by a factor of 16

Scenario	**Power/Area Increase/Decrease?**	**Numerical Factor**
You double the temperature.	(Increase) Decrease	**16**
You triple the temperature.	Increase Decrease	
You quadruple the temperature.	Increase Decrease	
You halve the temperature.	Increase Decrease	

5.8) The total power emitted by a given object will depend on both its temperature and its surface area – since the Stefan-Boltzmann law tells us the amount of power emitted from every individual square meter of the object's surface. Compare the **power per m²**, **surface area** and **total power emitted** by each of the objects described below. *(Choose greater than [>], equal to [=], or less than [<] in each column.)*

	Power/m²	**Surface Area**	**Total Power**	
2 m × 2 m × 2 m cube @ T = 10^4 K	> = <	> = <	> = <	2 m radius sphere @ T = 10^4 K
10 m radius sphere @ T = 10^3 K	> = <	> = <	> = <	10 m radius sphere @ T = 10^2 K
1 m × 1 m × 1 m cube @ T = 10^5 K	> = <	> = <	> = <	3 m × 3 m × 3 m cube @ T = 10^4 K
1 m radius sphere @ T = 10^4 K	> = <	> = <	> = <	4 m radius sphere @ T = 5×10^3 K

5.9) Consider the blackbody curve graphed below. Note that 1 nm = 10^{-9} m.

a) Label the y-axis of the blackbody graph (you don't need to add numerical values).

b) What is the approximate temperature (in K) of the blackbody whose curve is currently drawn in the graph?

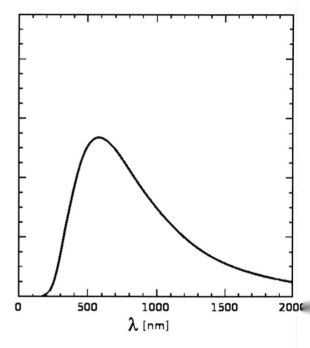

c) How much power per square meter does this object emit?

d) On the graph, draw the approximate curve of a blackbody with a temperature of 2,900 K. Be sure to label the curve.

e) On the graph, draw the approximate curve of a blackbody with a temperature of 14,500 K. Be sure to label the curve.

5.10) Observe the hydrogen lamp through the spectrometer.
   a) Draw in all the spectral lines that you observe below. Label the **color** of each line.

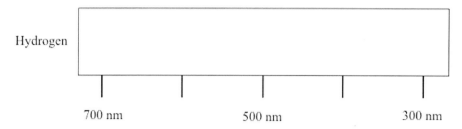

   b) Label the <u>precise</u> wavelength associated with each of the above hydrogen spectral lines (these values can be found in your textbook). If your measured values were significantly different than the actual values, what might have caused this error?

   c) In the energy level diagram below, draw in the transitions that have occurred to produce each of the above emission lines. Be sure to indicate the direction of each transition, and label the wavelength associated with each transition.

   ———— $n = \infty$
   ———— $n = 5$
   ———— $n = 4$

   ———— $n = 3$

   ———— $n = 2$

   ———— $n = 1$

   d) Why do you not observe any electron transitions that end in the n=1 level? Are these transitions not occurring? If they *are* occurring, why aren't they observable with this spectrometer?

   e) Why do you not observe any electron transitions that end in the n=3 (or higher) level? Are these transitions not occurring? If they *are* occurring, why aren't they observable with this spectrometer?

5.11) Examine the spectra[1] below. The bottom spectrum is a mixture of two or more of the elements above it. Which elements are present in the mixture *(circle each that is present)*?

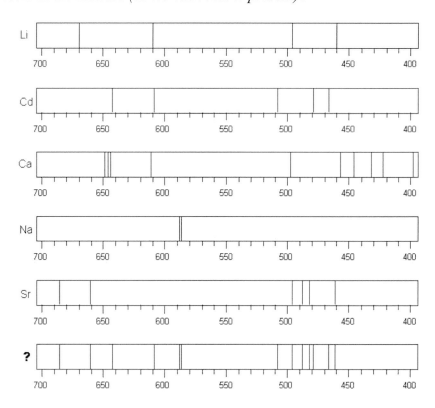

5.12) Light from each of the following sources is passed through a spectrometer (assume the spectrometer is sensitive to any relevant portion of the electromagnetic spectrum – not just optical light). Which of the following is observed: a **continuous**, **emission**, or **absorption** spectrum? *(In some cases, more than one type of spectrum may be observed simultaneously.)*

	Continuous	Emission	Absorption
Light emitted from a heated metal sphere			
Light emitted from the Orion nebula			
Light emitted from a human body			
Light emitted from an incandescent bulb's tungsten filament			
Light emitted from a florescent bulb			
Light emitted from the Crab nebula			
Light emitted from the Sun (or any star)			
Light emitted from a warm plastic cube			
Light emitted from Sun's corona (seen during an eclipse)			
Light from a planet (emitted light and reflected starlight)			

---

[1] Image from http://wikieducator.org/Chemistry/Light_Spectra

5.13) Later in this class we will be discussing **galaxies** in detail. Galaxies are complex systems composed of stars and gasses at many different densities and temperatures (cool gas, warm gas, hot gas, etc.). Would you expect a galaxy spectrum to be a **continuous**, **emission**, or **absorption** spectrum (or some combination thereof)?

5.14) The figure below shows the motion of five distant stars (A, B, C, D & E) relative to a stationary observer (telescope). The speed and direction of each star is indicated by the length and direction of the arrows shown.[2]

Observer   **Distant Stars**

Rank the Doppler shift of the light observed from each star from greatest "blueshift", through no shift, to greatest "redshift".

Greatest blueshift ____ ____ ____ ____ ____ Greatest redshift

Explain your reasoning:

---

[2] #5.14 & 5.15 adapted from Hudgins, Lee & Prather "Astronomy Interactives" (http://astro.unl.edu/interactives/)

5.15) The first spectrum shown below is of an element as it appears in a laboratory here on Earth. The spectra that follow are those of five stars (A - E) as seen from Earth.

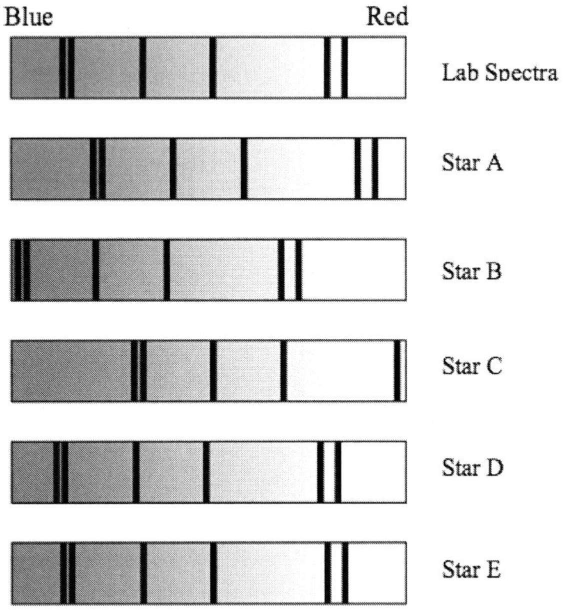

Rank the Doppler shift of the stars from greatest "blueshift", through no shift, to greatest "redshift".

Greatest blueshift ____ ____ ____ ____ ____ Greatest redshift

Which of the stars has some motion **toward** Earth? ____ ____
    Above, circle the star that has the faster motion towards Earth.

Which of the stars has some motion **away from** Earth? ____ ____
    Above, circle the star that has the faster motion away from Earth.

Which star is moving neither toward nor away from Earth? ____
Does the lack of observed Doppler shift indicate that this star is *completely stationary* in space? *Explain why or why not.*

If you could put a speedometer on each star, which would be moving the fastest through space? *Explain your reasoning:*

Chapter S4

S4.1) Fundamental particles.
 a) What is the difference between a **fermion** and a **boson**?

 b) What is the difference between a **quark** and a **lepton**?

 c) The most familiar particles are the electron, proton, and neutron as well as the neutrino and the photon. Indicate the properties of each of these particles below. <u>Cross out</u> any categories that are not applicable for that particle.

	Fermion or Boson?	Spin (Integer or Half-integer)?	Quarks or Lepton?	Charge?	Composed of which quarks?
Electron					
Proton					
Neutron					
Neutrino					
Photon					

S4.2) Antimatter!
   a) How much energy is released when a hydrogen atom and an anti-hydrogen atom meet and annihilate?

   b) If we could use anti-matter as an energy source, how many kilograms of anti-matter would be required to supply *all of mankind's energy needs for an entire year*? The human race currently uses approximately 5×10^{20} J/year. (Don't forget that <u>half</u> of the mass consumed in the annihilation process is just plain old regular matter.)

S4.3) Choose the correct fundamental force (**gravity**, **weak**, **electromagnetism**, or **strong**) associated with each statement.

	Gravity	Weak	EM	Strong
Which force is responsible for a neutron decaying into a proton (a process known as beta decay)?				
Which force bonds quarks together into particles like protons and neutrons?				
Which force governs the motion of an apple falling from a tree?				
Which force holds atomic nuclei together?				
What forces hold your cells together?				
What force governs the orbital patterns of planets around stars?				
What force is chemistry primarily concerned with?				
What force keeps stars and gas confined within a galaxy?				
What force is key to many biological processes such as protein synthesis and respiration/metabolism?				

S4.4) **Think like a scientist.** Make predictions about what would be observed in each of the following scenarios. Then test those predictions using a digital simulation. Note that the diagrams below are *not to scale – in particular, the width of the slits is highly exaggerated.*

a) **PREDICTIONS**: A focused light source is aimed at a screen. The light source emits light at just a single wavelength (like a laser). What pattern of light would you observe on the screen (sketch below in the left diagram)? Shade in the regions on the screen that would look brightest and leave blank those regions where little or no light would fall. If a thin, opaque plate pierced by two parallel slits was placed in between the light source and the screen, how would the pattern on the screen change (sketch below in the right diagram)?

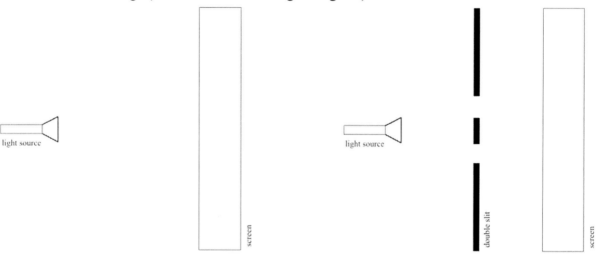

Explain your predictions. Why do you think the above patterns will occur?

b) **OBSERVATIONS:** Launch the Quantum Wave Interference digital simulation. Leave all the settings on the default and turn on the laser. Sketch the pattern of light you observe on the screen (in the left diagram). Stop the laser and clear the screen. Then, select the "Double Slits >>" button in the simulation. This will add a double slit to the experiment (leave all the settings on the default). Sketch the pattern of light you observe on the screen (in the right diagram).

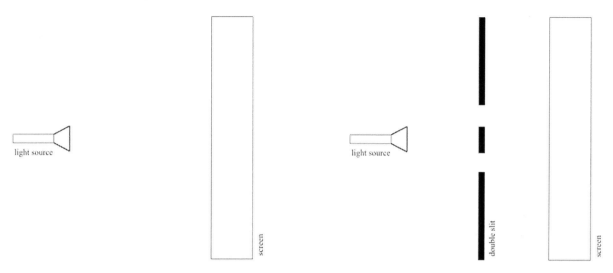

Observations consistent with predictions?
YES NO   *(circle one)*

Observations consistent with predictions?
YES NO   *(circle one)*

If your observations do not match your predictions, are you surprised or do the observations (in hindsight) make sense? Can you explain what you observe?

c) The pattern that arises when light passes through the double slit is a result of light waves constructively and destructively interfering with one another. Constructive interference occurs when wave crests meet. Destructive interference occurs when a wave crest meets a trough. What type of interference is occurring at each region along the screen (*circle either constructive or destructive in each row* – note that in this diagram light areas are shown in white and dark areas are shown in black)?

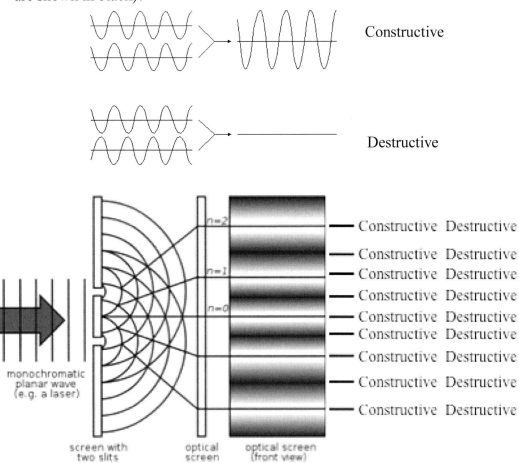

Constructive

Destructive

— Constructive Destructive
— Constructive Destructive
— Constructive Destructive
— Constructive Destructive
— Constructive Destructive
— Constructive Destructive
— Constructive Destructive
— Constructive Destructive
— Constructive Destructive

d) **PREDICTIONS:** An electron "gun" is aimed at a screen. <u>This electron source releases one electron at a time.</u> Each time an electron hits the screen, the screen is illuminated at the site of impact. What pattern of light would you observe on the screen if you ran your experiment for an extended period of time so that you captured many electron hits (sketch below in the left diagram)? If a thin, opaque plate pierced by two parallel slits were placed in between the electron source and the screen, how would the pattern on the screen change (sketch below in the right diagram)?

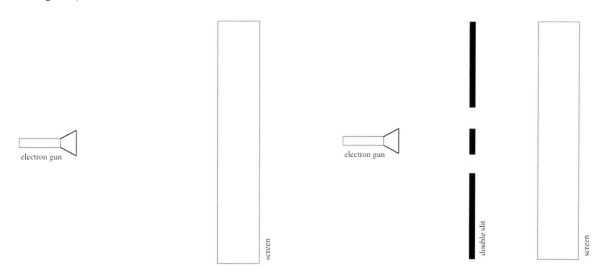

e) **OBSERVATIONS:** Select the "Single Particles" tab. Check "Auto-Repeat," check "Rapid," and <u>un</u>-check "Fade." Leave all the other settings on the default and turn on the electron gun. ***Allow the simulation to run for several minutes***, then sketch the pattern of electron hits you observe on the screen (in the left diagram). Stop the electron gun and clear the screen. Then, select the "Double Slits >>" button in the simulation. This will add a double slit to the experiment (leave all the settings on the default). ***Allow the simulation to run for several minutes***, then sketch the pattern of light you observe on the screen (in the right diagram).

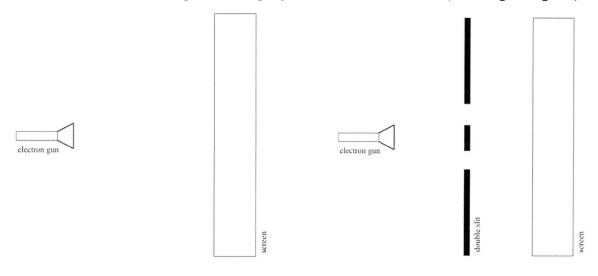

Observations consistent with predictions?　　　Observations consistent with predictions?
YES　NO　*(circle one)*　　　　　　　　　　YES　NO　*(circle one)*

f) What can you conclude about the nature of electrons based on the electron double-slit experiment?

g) When you send a single electron at a time toward the screen (with no slits in the way), why doesn't each electron simply hit in the same place on the screen again and again? (This is also why the simulation represents each electron emerging from the gun as a large fuzzy patch, rather than a small, localized particle).

h) In the double-slit experiment carried out with light, there are obviously multiple waves of light that can interact constructively and destructively, leading to the interference pattern on the screen. But in the electron double-slit experiment, only 1 single electron was emitted at a time. What, then, is each electron interacting with to produce the interference pattern?

S4.5) In "Star Trek" the transporter is used to move objects and people from place to place almost instantaneously. The transporter is said to disassemble matter down to the subatomic level, and then reassemble it elsewhere in its exact original state. To accomplish this feat, engineers from the future make use of a (fictitious) device called the "Heisenberg Compensator." What might such a device be "compensating" for? Why would such a device be necessary for a Star Trek-like transporter system to work?

S4.6) What type of pressure corresponds to each of the following statements? *(Select all that apply in each row.)*

	Thermal Pressure	Electron Degeneracy Pressure	Neutron Degeneracy Pressure
Pressure found at only the very highest densities of material.			
The pressure that keeps a party balloon inflated.			
The pressure that keeps a neutron star from collapsing into a black hole (at least until the speed of the neutrons approaches the speed of light – at around 3 M_{Sun}).			
The type of pressure that depends on temperature.			
The type of pressure that does not depend on temperature.			
The pressure that supports a white dwarf (at least until it reach the *Chandrasekhar Limit* of ~ 1.4 M_{Sun}).			
The pressure that drives the pistons in a typical car engine.			
Pressure associated with low-density material.			
The pressure that keeps a brown dwarf from igniting hydrogen fusion and becoming a star.			
Resistance to compression that stems from the exclusion principle.			
Pressure only found in material composed purely of neutrons.			

S4.7) **Think like a scientist.** You come across a website on which an engineer from a respected academic institution claims that she has discovered a way to harness zero point energy (vacuum energy). She states that this device could fulfill all of humankind's energy needs for the rest of time at basically no cost. She has not published this work in any of the predominant, peer-reviewed physics or engineering research journals.

   a) List and explain at least 2 reasons that you should be skeptical of these claims.

   b) Imagine that you had the opportunity to meet this engineer. Make a list of questions that you would ask her in order for you to evaluate – in a scientific manner – the veracity of her claims.

Chapter 14

14.1) The stellar interior:
   a) Label each of the Sun's layers in the diagram below.

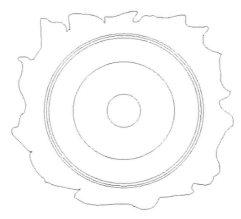

   b) List the layers of the Sun in order of **temperature**.

   Hottest: _____

   _____

   _____

   _____

   _____

   Coolest: _____

   c) List the layers of the Sun in order of **density** (# of particles per area).

   Most dense: _____

   _____

   _____

   _____

   _____

   Least dense: _____

   d) How are temperature and density related in the Sun?

14.2) The surface of the Sun has an average temperature of 5800 K and a radius of 6.95×10^8 m. Based on this information, solve the following problems. Recall that power is measured in the units of Watts, where 1 Watt = 1 Joule/second.

   a) How much power does the Sun emit per square meter of surface area? Your answer should have units of W/m².

   b) How much total power does the Sun emit from its entire surface? This value is the Sun's **luminosity**. (After calculating, check your value against that quoted in the textbook.)

   c) If the Sun shines with the same luminosity over its entire 10 billion year life span, then how much mass does the Sun convert into energy during its lifetime? Recall that $E=mc^2$.

   d) Given that the mass of the Sun is about 2×10^{30} kg, what percentage of the solar mass is lost in this energy conversion during the Sun's lifetime?

   e) How many kilograms of hydrogen must be fused into helium to power the Sun over its entire lifetime? Recall that in hydrogen fusion only 0.7% of the initial hydrogen mass is converted into energy. (The remaining 99.3% becomes helium).

   f) What percent of the Sun's mass is converted from H → He over its lifetime?

14.3) Draw the 3 steps of the **proton-proton chain** below. Label each particle involved in the chain.

14.4) While scientists in the 1920s believed that the Sun must be powered by hydrogen fusion, they recognized that there was a major problem with the theory: the temperature in the interior of stars was not nearly high enough for proton-proton fusion to actually occur! This is because the electromagnetic repulsive force between two positive charged protons is incredibly strong relative to the kinetic energy of protons in the core of the Sun. Even at 15 million K, scientists calculated that protons could almost never get close enough for the strong force to overcome this repulsion. By the 1940s, scientists (including George Gamow, who was a professor at GW from 1934 to 1954) had solved the problem. What major discovery "fixed" the problem? Explain why this discovery allows fusion to occur in the core of the Sun.

14.5) Imagine that the laws of physics suddenly changed and the strong force that binds nuclei together was suddenly 100 times weaker. What would happen to the rate of fusion in the core of the Sun? Would it increase, decrease, or remain the same? *Explain.*

14.6) While the Sun remains primarily in a state of **hydrostatic equilibrium** through its lifetime, it does undergo some changes at it ages.
   a) As the Sun fuses hydrogen into helium, what happens to the number of individual particles in the core of the Sun? Does the number increase or decrease? *Explain why.*

   b) How does the Sun's core react to the change in the number of individual particles in its core?
      i. Does the core size increase or decrease? *Explain why.*

      ii. Does the core temperature increase or decrease? *Explain why.*

      iii. Does the rate of fusion in the core increase or decrease? *Explain why.*

   c) As a result of the above changes, what happens to the solar luminosity over a long period of time (i.e. billions of years)?

14.7) Astronomers have never sent a probe into the interior of the Sun. How, then, do astronomers know about the Sun's interior structure?

14.8) Imagine that the laws of physics suddenly changed, and fusion could no longer happen in the core of the Sun (e.g. the fusion reaction suddenly ceases and <u>cannot</u> be restarted!).
    a) Would the Sun immediately stop shining? Why or why not?

    b) How would astronomers know that the Sun's fusion reactions had stopped?

14.9) **Think like a scientist.** You are aware that at some point in the future the Sun will run out of hydrogen fuel at its core and begin to "die." Based on what you know about the properties of matter and the interaction of pressure and gravity in the interior of the Sun, make hypotheses about what will happen in the core of the Sun when it runs out of hydrogen and *explain your reasoning.*

    a) What will happen to the size of the Sun's core when it runs out of hydrogen fuel? Will it get bigger, smaller, or remain the same size?

    b) If the size of the Sun's core changes when it runs out of hydrogen fuel, what limits might there be on how large or how small the core becomes?

    c) If the size of the Sun's core changes after it runs out of hydrogen fuel, what will happen to the core temperature of the Sun? Will the temperature go up, down, or remain the same?

    d) How might you test your hypotheses if you had access to any resources (telescopes, computers, preexisting data, etc.) that you desire? (Note that you are limited to current, real world knowledge and technology.)

Chapter 15

15.1) Consider the equation for the inverse square law for light:

$$b = \frac{L}{4\pi d^2}$$

a) Imagine that you are observing a light source. How would the apparent brightness of the light source change if the light source (or the distance to the light source) was altered in the following ways? Indicate if the apparent brightness would increase or decrease, and calculate the factor by which it would change. The first row has been completed for you as an example.

Example: You double the luminosity of the light source.

$$b(Initial) = \frac{L}{4\pi d^2}$$

$$b(Final) = \frac{2 \times L}{4\pi d^2} = 2 \times \frac{L}{4\pi d^2} = 2 \times b(Initial)$$

therefore $b(Final) = 2 \times b(Initial)$
→ the apparent brightness has increased by a factor of 2

Scenario	Apparent brightness Increase/Decrease?	Numerical Factor
You double the luminosity of the light source.	(Increase) Decrease	2
You triple the luminosity of the light source.	Increase Decrease	
You halve the luminosity of the light source.	Increase Decrease	
You double the distance to the light source.	Increase Decrease	
You triple the distance to the light source.	Increase Decrease	
You halve the distance to the light source.	Increase Decrease	
You double the luminosity of the light source while also doubling the distance to the light source.	Increase Decrease	
You halve the luminosity of the light source while also halving the distance to the light source.	Increase Decrease	
You quadruple the luminosity of the light source while also halving the distance to the light source.	Increase Decrease	
You quadruple the luminosity of the light source while also doubling the distance to the light source.	Increase Decrease	

b) Calculate the distance to a star that is half as luminous as the Sun but has an apparent brightness of just 1.1×10^{-8} W/m². (For reference, the apparent brightness of the Sun is 1.4×10^{3} W/m² and the apparent magnitude of Vega, a relatively bright star in the night sky, is 2.7×10^{-8} W/m².)

1) Identify which variables are known (given) and which are unknown (to be solved for). Write down each known value *including units*.

    KNOWN VARIABLES: UNKNOWN VARIABLES:

2) Convert all known values into the units required by the equation. In the case of the equation for the inverse square law for light, the units that must be used are Watts and meters – we must stick to these units because apparent brightness is measured in W/m² (recall that a Watt is a Joule/second – a unit of power).

    b in W/m² = _____

    L in W = _____

3) **Algebraically** solve the equation for the unknown value **before** plugging in any numbers. In this case, the unknown value is the distance to the star. Show your work below.

    $d =$

4) Finally, use your calculator to solve for the distance. Be careful with your order of operations, and don't forget any of the exponents!

    d (distance to the star) = _____

To get a better sense of how far this is, convert the distance into light years:

    distance to the star in light years = _____

15.2) Consider the diagram above.[3]

a) In October, you are observing the night sky. Use a ruler to draw a straight line from Earth in October, through the "Nearby Star", out to the "Distant Stars." In the image of the sky shown to the right, draw the position of the "Nearby Star" as viewed in October.

b) Repeat the 'part a' procedure for the "Nearby Star" as viewed in January.

c) Repeat the 'part a' procedure for the "Nearby Star" as viewed in July.

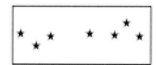

d) Describe how the "Nearby Star" appears to move in the night sky over the course of a year. This motion is called **stellar parallax**.

e) In the diagram, draw a new star between Earth and the "Nearby Star". During the course of a year, will this closer star appear to move back and forth through the night sky more or less than the "Nearby Star"? Why?

f) In the diagram, draw a new star between the "Nearby Star" and the "Distant Stars". During the course of a year, will this further star appear to move back and forth through the night sky more or less than the "Nearby Star"? Why?

---

[3] #15.2 & 15.3 adapted from Prather et al. "Lecture Tutorials for Introductory Astronomy"

15.3) Consider the stars field shown in the figure to the right. This represents a tiny patch of the night sky. In this image the angle separating Stars A and B is just ½ of an arcsecond.

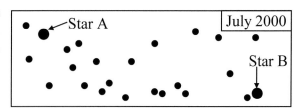

a) In the figure on the next page there are images of this star field taken at different times during the year. One star in the field exhibits parallax with respect to the other, more distant stars. Determine which star exhibits parallax, and circle that star in *each* image on the next page. In the image above, draw a line that shows the range of motion for the nearby star. Label the endpoints of this line with the months when the star appears at those points.

b) What is the angular separation between the endpoints that you have marked in the above image? (Use the known separation between Stars A and B to estimate this value.)

c) What is the <u>parallax angle</u> for the nearby star? Don't forget that a star's parallax is equal to **half** the star's annual back-and-forth shift in position.

d) What is the distance to this nearby star, in units of both *parsecs* and *light years*?

e) If the nearby star was twice as far away as what you calculated in 'part d', what would its parallax angle be?

f) If the nearby star was at half the distance you calculated in 'part d', what would its parallax angle be?

g) Describe *in words* how a star's parallax angle is related to its distance. In particular, how does the parallax angle change as distance increases?

h) Would very distance stars exhibit large or small back-and-forth shifts over the course of a year?
   LARGE   SMALL   *(circle one)*

i) Would very nearby stars exhibit large or small back-and-forth shifts over the course of a year?
   LARGE   SMALL   *(circle one)*

j) Would you expect **galaxies** to exhibit any measureable parallax? *Explain why or why not.*

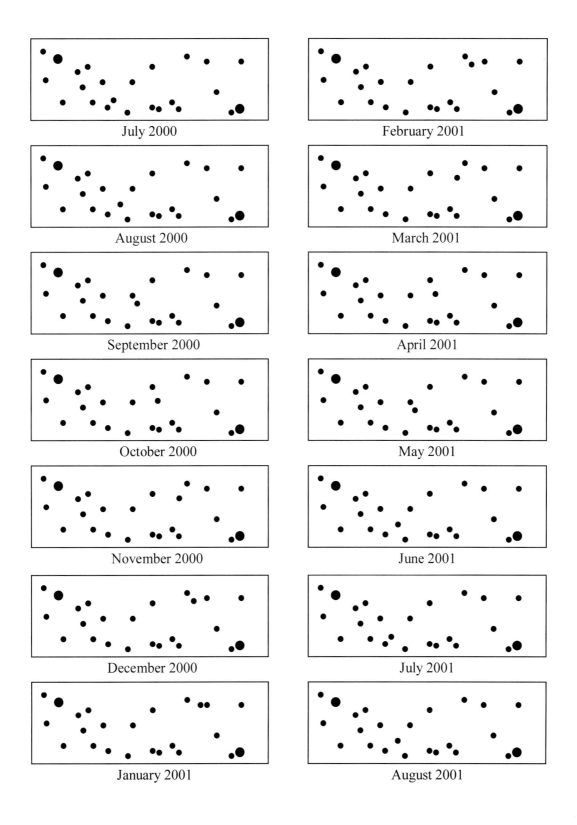

15.4) Stellar spectral type, luminosity class and the H-R diagram.
   a) How is a star's spectral type determined? Be specific.

   b) Why do spectra of the hottest stars (O & B) exhibit very few spectral lines?

   c) Why do spectra of the coolest stars (K & M) exhibit very many spectral lines?

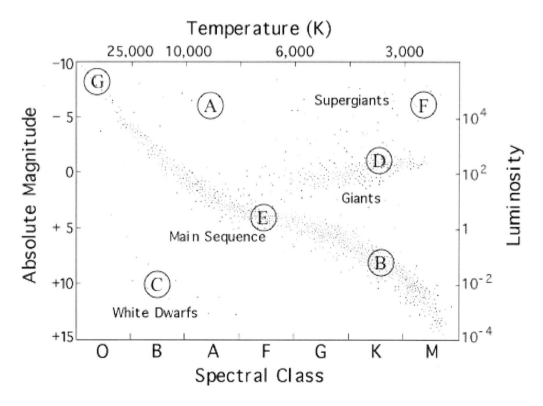

   d) What is the approximate spectral type and luminosity class of the stars at the positions labeled on the H-R diagram above?

	A	B	C	D	E	F	G
Spectral type							
Luminosity class							

e) Answer each question below by circling the letter (or letters) corresponding to the correct star (or stars) in the H-R diagram on the previous page.

Which star is the hottest?	A B C D E F G
Which star is the coolest?	A B C D E F G
Which star is the largest?	A B C D E F G
Which star is the smallest?	A B C D E F G
Which star is the reddest in color?	A B C D E F G
Which star is the bluest in color?	A B C D E F G
Which two stars have the same temperature?	A B C D E F G
Which two stars have the same luminosity?	A B C D E F G
Which stars fall on the main sequence?	A B C D E F G
Of the stars on the main sequence, which will live the longest amount of time?	A B C D E F G
Of the stars on the main sequence, which will live the shortest amount of time?	A B C D E F G
Which stars have already used up all the hydrogen fuel in their cores?	A B C D E F G
In which star has fusion stopped entirely?	A B C D E F G

f) Explain why Star F must be larger in radius than Star A.

g) Explain why Star D must be larger in radius than Star B.

h) There appear to be some regions of the H-R diagram where no stars are plotted on the graph. Why might these regions be empty? Do no stars exist with those combinations of temperatures and luminosities – if so, why not? Or are there other explanations?

15.5) The mass of *most* stars cannot be measured directly (notice, for example, that no masses are listed in your textbook's "Stellar Data" appendix). This is unfortunate, since mass is a star's most fundamental property! How then do astronomers estimate the mass of stars?

a) The mass of *some* stars can be measured. What very specific scenario is required in order for astronomers to be able to accurately calculate stellar mass using Newton's Version of Kepler's Third Law?

b) Given any random star, how can you estimate its mass? What specific type of observations of this star would you require? What other data/references would be needed?

15.6) Imagine that you are observing and categorizing stars. One of the stars you observe has a spectral type and luminosity that makes it fall within the instability strip on the H-R diagram. What observations would you require in order to test whether or not you have identified a variable star?

15.7) Star clusters are extremely useful tools for astronomers.
   a) Compare the two types of clusters:

	Number of stars	Size	Age	Location
Open clusters				
Globular clusters				

b) What are the two key reasons that star clusters are so important to astronomers?

#1)

#2)

c) The H-R diagrams[4] of four different star clusters are shown below. In each of the diagrams below, circle the **main sequence turnoff point.**

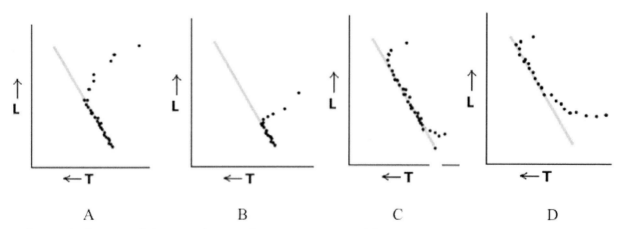

A   B   C   D

d) Rank the age of the star clusters from youngest to oldest.

Youngest ____ ____ ____ ____ Oldest

Explain your reasoning:

---

[4] *Adapted from Hudgins, Lee & Prather "Astronomy Interactives" (http://astro.unl.edu/interactives/)*

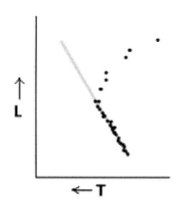

e) Imagine that the main sequence turnoff point for the above star cluster is around the spectral type A. In order to estimate the age of this cluster, what additional piece of information do you need?

f) Could the star cluster shown above contain any main-sequence B stars? *Explain why or why not.*

g) Could the star cluster shown above contain any main-sequence K stars? *Explain why or why not.*

Chapter 16

16.1) Imagine a light source that emits an equal amount of light at every wavelength. The spectrum of this light source is shown below.

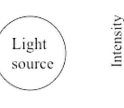

a) Now imagine this same light source is seen through a typical cloud of interstellar dust and gas. How would the observed spectrum of this light source change? (Note that the intrinsic spectrum of the light source remains unchanged.)

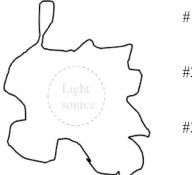

#1.

#2.

#3.

b) Sketch in an approximation of the new spectrum (the original spectrum is shown as a guideline). Be sure that all 3 changes indicated in 'part a' are reflected in the graph.

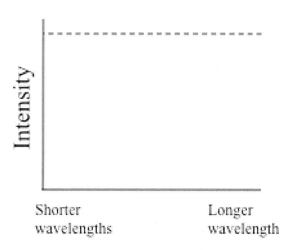

$$M_{balance} = 18 M_{Sun} \times \sqrt{\frac{T^3}{n}}$$

and if $M_{molecular\ cloud} \geq M_{balance}$, then star formation can occur!

16.2) A molecular cloud in space has a mass of 100 M_{Sun}. The cloud has a temperature of 10 K and an average density of 200 particles per cubic centimeter.
  a) Given the right impetus (such as the explosion of a nearby star), could this molecular cloud form stars? *Why or why not?*

  b) How <u>dense</u> must a molecular cloud be in order for star formation to proceed if the molecular cloud has a mass of 100 M_{Sun} and a temperature of 10 K?

  c) What <u>temperature</u> would be required for star formation to proceed if the molecular cloud has a mass of 100 M_{Sun} and density of 200 particles/cm^3?

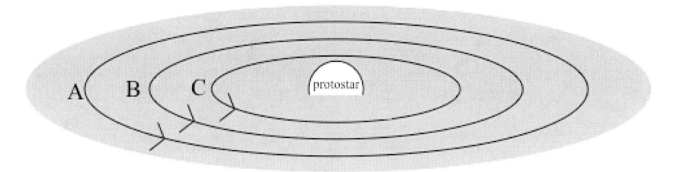

16.3) Above is a very rough sketch of an accretion disk (seen partially edge-on). *A real accretion disk would have thickness and would most likely not have uniform density, nor would it have such a well-defined outer edge.* The orbital paths of three particles in the disk are indicated (A, B & C).

   a) Rank the orbital speed of each particle: Slowest ___ ___ ___ Fastest
      Explain you reasoning:

   b) Rank the temperature in the disk near each particle: Coolest ___ ___ ___ Hottest
      Explain you reasoning:

   c) If particles A, B and C were the only particles orbiting the protostar, then their orbital parameters (period and semi-major axis) would remain constant over time because of the conservation of _____ and _____. In an accretion disk, however, outer particles will spiral inward. Explain why this is possible in an accretion disk, and why this process does not violate either conservation law.

   d) In the image above, sketch in protostellar jets.

16.4) On the H-R diagram below, sketch the approximate pre-main sequence life tracks of stars with the following main sequence masses: 15 M_{Sun}, 3 M_{Sun}, 1 M_{Sun} and 0.5 M_{Sun}.

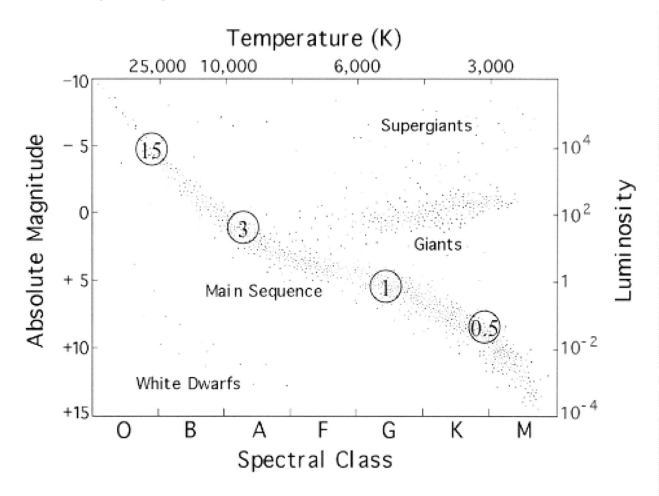

a) How long does each type of protostar take to become a main sequence star? Indicate that time on the H-R diagram near each protostar's life track.

b) Notice that the life track of the 15 M_{Sun} star crosses through the supergiant region of the H-R diagram. Supergiant stars are actually at the end of their lives, while supergiant protostars are just beginning their lives. Can astronomers distinguish between these two types of objects?

16.5) What is the lowest mass that a star might have? What is the highest mass that a star might have? *Explain the reason for each limit.*

16.6) Low mass stars are much more common than high mass stars. In the Milky Way, for example, for every 250 low-mass (< 2 M_{Sun}) stars that form just 1 single high-mass (>10 M_{Sun}) star forms! Additionally, recall that high-mass stars (lifetime ~ millions of years) die very quickly compared to low-mass stars (lifetime ~ billions of years).
   a) Given how quickly high-mass stars in the Milky Way burn out, how can astronomers determine the relative number of high-mass (>10 M_{Sun}) vs. low-mass (< 2 M_{Sun}) stars that initially form?

b) Imagine that a large molecular cloud in the Milky Way undergoes a burst of star formation, forming a cluster of approximately 1,000 new stars. Would most of the mass of these stars be found in high-mass (>10 M_{Sun}) or low-mass (<2 M_{Sun}) stars? *To answer this problem, estimate and then compare the total amount of mass in each population. Explain what assumptions you make in your calculation.*

c) Imagine that a large molecular cloud in the Milky Way undergoes a burst of star formation, forming a cluster of approximately 1,000 new stars. Would most of the mass of these stars be found in high-mass (>10 M_{Sun}) or low-mass (<2 M_{Sun}) stars? *To answer this problem, estimate and then compare the total amount of mass in each population. Explain what assumptions you make in your calculation.*

Chapter 17

17.1) The most counterintuitive stage in the death of any star is the simultaneous contraction of the core and expansion of the outer layers of the star after the star has run out of hydrogen fuel in its core. Explain *in detail* and in your own words why/how the star swells into a red giant or red supergiant.

17.2) Immediately after a low-mass star moves off the main sequence, it moves up and to the right in the H-R diagram. How are the star's temperature, size and luminosity changing during this period of time? *Explain why these changes are occurring.*

17.3) **Think like a scientist.** The textbook includes a diagram of the **triple alpha reaction** that shows three helium-4 nuclei coming together to form one carbon-12 nucleus. This seems to imply that the three helium-4 nuclei must collide simultaneously for this fusion reaction to occur. Given what you know about the size of atomic nuclei and the process of fusion (including electromagnetic repulsion, the weak force, and quantum tunneling) do you think that the triple alpha reaction is as simple as the diagram implies (e.g. just one step)? Is it likely that three helium-4 nuclei can collide simultaneously? *Explain your reasoning*. If you think that an intermediate step (or steps) is required in this reaction, what might that step (or steps) be? [Note that the Periodic Table of Elements is available in the appendix of your textbook.]

17.4) After the **helium flash** in a low-mass star, it moves down and to the left in the H-R diagram. How are the star's temperature, size and luminosity changing during this period of time? Explain why these changes are occurring.

17.5) In the H-R diagram below, sketch the approximate life track of a 1 M_{Sun} star after it turns off the main sequence to its final state as a white dwarf. Label the region (or regions) in which the star is undergoing **hydrogen shell fusion, helium core fusion** and **helium shell fusion**. Label the point where the **helium flash** occurs. Indicate with a dashed line the transition between the red giant and white dwarf phases of this star – during this time the star's outer layers are ejected into a planetary nebula.

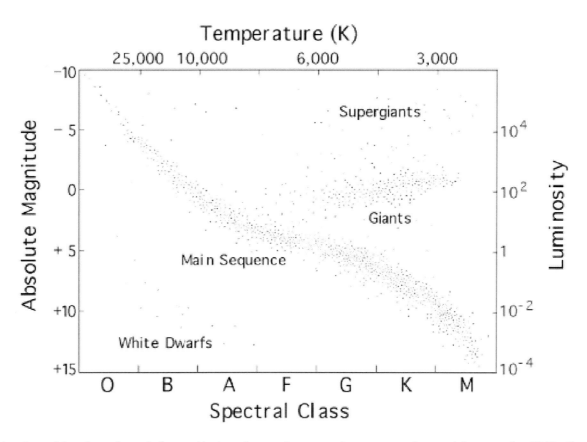

17.6) As the white dwarf cools by radiating thermal energy into space, its position on the H-R diagram moves down and to the right. During this cooling, does the radius of the white dwarf change? *Explain why or why not.* Check your answer by consulting the extremely detailed, full-page version of the H-R diagram in Chapter 15 of your textbook.

17.7) When a low-mass star ejects its outer layers, this gas forms a **planetary nebula** (so named because these nebulae look deceptively like planets through very small telescopes).
   a) Most of the helium and carbon (and a bit of nitrogen and oxygen) produced during the main-sequence lifetime of a low-mass star will remain forever trapped in the white dwarf, but some of this material gets mixed into a star's outer hydrogen layers and ejected in the planetary nebula. With each generation of stars, these heavier elements "enrich" the galaxy's interstellar medium. If the universe contained *only* low-mass stars, could humans (or similar beings) exist? After all, our bodies are primarily composed of carbon and water (H_2O). *Explain your reasoning.*

   b) What type of spectrum would you expect planetary nebulae to produce: emission, absorption or continuous? *Explain your reasoning.*

   c) The number of planetary nebulae observed in the Galaxy is well under 2,000. Even considering our inability to detect dim planetary nebulae or those blocked by dust in the Galactic plane, there are relatively few planetary nebulae in the Galaxy given the number of low-mass stars. What must happen to planetary nebulae over time?

17.8) Why is the **CNO cycle** only possible in the cores of high-mass stars? How does the rate of fusion in the CNO cycle compare to the rate of fusion in the proton-proton chain?

17.9) In the H-R diagram below, sketch the approximate life track of a 10 M_{Sun} star after it turns off the main sequence.

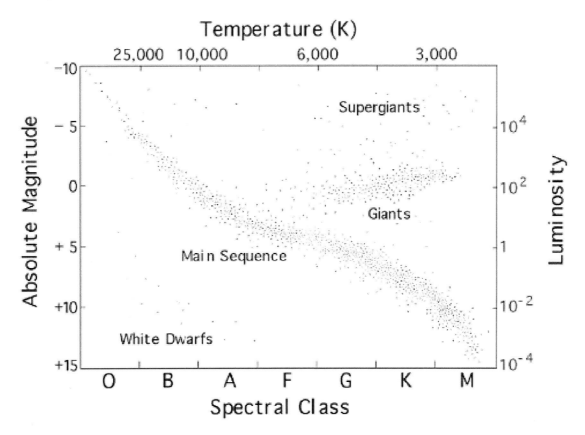

17.10) The position of high-mass stars after they leave the main sequence basically only shifts horizontally through the H-R diagram. This means that these stars change in _____ while their _____ remains constant. What is happening in the core of high-mass stars during this period of time? How does the shift remain purely horizontal?

17.11) The core of a high-mass star may become a neutron star after the star goes supernova. White dwarfs are plotted on the H-R diagram, so why aren't neutron stars? (Black holes obviously can't be plotted since, by definition, they emit no light!)

17.12) Why does it take a supernova explosion to produce elements heavier than iron?

17.13) Once gravity overcomes electron degeneracy pressure, electrons and protons are forced together to produce neutrons (and hence a neutron star). In this process of "electron capture," neutrinos are produced. What role do these neutrinos play in supernovae?

17.14) Answer each question below by circling the letter corresponding to the correct star (or stars) in the H-R diagram.

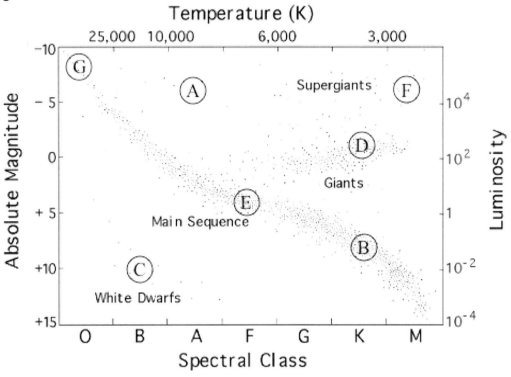

Which stars are undergoing hydrogen fusion in their cores?	A B C D E F G
Which stars are undergoing any type of hydrogen fusion?	A B C D E F G
Which stars are undergoing any type of helium fusion?	A B C D E F G
Which stars might be undergoing fusion of elements heavier than helium?	A B C D E F G
Which star is likely the oldest?	A B C D E F G
Which star is likely the youngest?	A B C D E F G
Which star on the main sequence is likely the oldest?	A B C D E F G
Which star on the main sequence is likely the youngest?	A B C D E F G
Which stars are likely to explode as supernovae?	A B C D E F G
Which star is most likely to be the next star to go supernova?	A B C D E F G
Which stars are likely to end up as white dwarfs?	A B C D E F G
Which star is most likely to be the next star to become a white dwarf?	A B C D E F G
Which stars will likely produce planetary nebulae?	A B C D E F G
Which stars will produce iron cores?	A B C D E F G
Which stars will likely become either neutron stars or black holes?	A B C D E F G
Which stars have experienced or will experience a helium flash?	A B C D E F G
Which stars have "onion-like" layers of shell fusion around their core?	A B C D E F G
Which stars will be responsible for "polluting" the interstellar medium with heavy elements?	A B C D E F G

Chapter S2

S2.1) Are the following pairs of observers in the same reference frame or in different reference frames?

	Same frame	Different frame
Two observers standing next to each other on the surface of the Earth.		
Two observers in the same spaceship.		
Two observers in different spaceships. One ship is trailing the other ship by a few kilometers, but travelling at the same speed.		
Two observers in different spaceships. One ship is trailing the other ship by a few kilometers, but travelling at a slightly different speed.		
Two observers in different spaceships. The ships are heading toward each other at the same speed (don't worry, they'll turn before a collision occurs!).		
One observer in a spaceship orbiting the Earth, the other observer standing on the surface of the Earth.		
One observer on Earth, the other observer on Mars.		

S2.2) List the two fundamental tenets of the theory of relativity:

#1

#2

S2.3) Special relativity is special because it is applicable only for the "special case" of objects moving in the absence of acceleration (including gravitational fields). Therefore, we are limited to considering only **inertial reference frames** (or "free-floating" frames).

a) You are in a <u>windowless</u> spaceship (the box sketched below). Imagine that in your spaceship there is a square object held motionless by a retractable rod. If the rod were retracted in each of the following scenarios, what would happen to the square object (e.g. describe its motion)? Would you feel weighted or weightless in each scenario? Which spaceship (or ships) is considered to be an *inertial* frame of reference?

Scenario A: The spaceship is parked on the surface of the Earth.

Behavior of square object:

Your weight: *(circle one)* Weighted or Weightless

Inertial reference frame?: *(circle one)* YES or NO

Scenario B: The spaceship is traveling through deep space at a constant velocity (far from any massive objects like stars or planets).

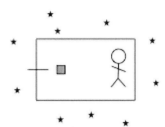

Behavior of square object:

Your weight: *(circle one)* Weighted or Weightless

Inertial reference frame?: *(circle one)* YES or NO

Scenario C: The spaceship is traveling through deep space firing its rockets in order to accelerate at a rate of approximately 9.8 m/s^2.

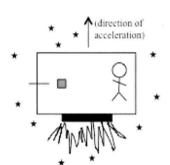

Behavior of square object:

Your weight: *(circle one)* Weighted or Weightless

Inertial reference frame?: *(circle one)* YES or NO

b) Given that you cannot see anything outside the walls of your spacecraft, can you tell the difference between scenario A (gravity) and C (acceleration) or are they **equivalent**? Would you expect the outcome of a physics experiment conducted in scenario A to be the same or different than if the identical experiment was conducted in scenario C?

S2.4) You are in the inertial (free-floating) frame of you own personal spaceship.
   a) If you didn't look at your ship's speedometer or out the window, how fast would you think you were moving?

You look out your window and observe another spaceship pass by at 100 m/s. Inside that spaceship, astronaut Bob throws a baseball at 40 m/s (as diagrammed below).

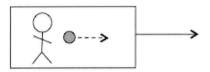

b) How fast do you perceive astronaut Bob to be moving relative to yourself? _____

c) How fast does astronaut Bob perceive himself to be moving? _____

d) How fast does astronaut Bob perceive you to be moving? _____

e) How fast do you perceive the baseball to be moving relative to yourself? _____

f) How fast does astronaut Bob perceive the baseball to be moving? _____

g) How fast does the baseball perceive you to be moving? _____

You look out your window and observe another spaceship pass by at 0.9c. Inside that spaceship, astronaut Jack turns on a flashlight (as diagrammed above).

h) How fast do you perceive astronaut Jack to be moving relative to yourself? _____

i) How fast does astronaut Jack perceive himself to be moving? _____

j) How fast does astronaut Jack perceive you to be moving? _____

k) How fast do you perceive the light to be moving relative to yourself? _____

l) How fast does astronaut Jack perceive the light to be moving? _____

S2.5)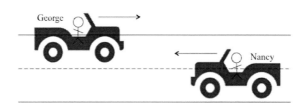

George and Nancy are passing each other on the highway. George is travelling at 100 m/s. Nancy is travelling at 75 m/s.

a) How fast does George perceive Nancy to be travelling? _____

b) How fast does Nancy perceive George to be travelling? _____

Jane and Sam are passing each other in their spaceships. Jane is travelling at 0.8c. Sam is travelling at 0.7c.

c) How fast does Jane perceive Sam to be travelling? *(Circle one choice below.)*

Less than 0.7c	0.7c	Greater than 0.7c but less than 0.8c	0.8c	Greater than 0.8c but less than c	c	1.5c

d) How fast does Sam perceive Jane to be travelling? *(Circle one choice below.)*

Less than 0.7c	0.7c	Greater than 0.7c but less than 0.8c	0.8c	Greater than 0.8c but less than c	c	1.5c

e) If Jane was to shine a flashlight toward Sam, how fast would the light be travelling according to Sam? _____

f) Imagine that the two ships pass by one another. The distance between the two ships is now increasing. If Jane was to shine a flashlight toward Sam's receding ship, how fast would the light be travelling according to Sam? _____

g) Is there any direction or speed that Jane could travel such that the light from the flashlight in her ship would appear to Sam to be moving at a speed of greater than the speed of light? *Explain your reasoning.*

S2.6) You and a friend meet and compare spaceships and clocks. Both spaceships are the same length and the same mass. Your clocks are synchronized and running at the exact same rate. You head off into space together.
  a) You start by traveling side-by-side at a speed of 0.75c. What would you and your friend observe? *(Circle one phrase in each row to complete the sentence.)*

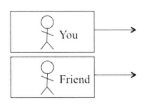

Traveling side-by-side at a speed of 0.75c.		
You observe the length of your own spaceship to be	longer than / shorter than / the same as	when it was at rest.
You observe the clock in your own ship to be running	faster than / slower than / the same as	when it was at rest.
You observe the mass of your own ship to be	more massive than / less massive than / the same as	when it was at rest.
You observe the length of your friend's ship to be	longer than / shorter than / the same as	your spaceship.
You observe the clock in your friend's ship to be running	faster than / slower than / the same as	your clock.
You observe the mass of your friend's ship to be	more massive than / less massive than / the same as	your spaceship.
Your friend observes the length of your spaceship to be	longer than / shorter than / the same as	his own spaceship.
Your friend observes the clock in your ship to be running	faster than / slower than / the same as	his own clock.
Your friend observes the mass of your ship to be	more massive than / less massive than / the same as	his own ship.

b) If, instead, you allowed your friend's spaceship to pass by you at a relative speed of 0.9c, what would you and your friend observe? *(Circle one phrase in each row to complete the sentence.)*

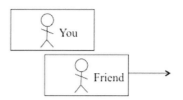

Friend passes by you at a relative speed of 0.9c.		
You observe the length of your own spaceship to be	longer than / shorter than / the same as	when it was at rest.
You observe the clock in your own ship to be running	faster than / slower than / the same as	when it was at rest.
You observe the mass of your own ship to be	more massive than / less massive than / the same as	when it was at rest.
You observe the length of your friend's ship to be	longer than / shorter than / the same as	your spaceship.
You observe the clock in your friend's ship to be running	faster than / slower than / the same as	your clock.
You observe the mass of your friend's ship to be	more massive than / less massive than / the same as	your spaceship.
Your friend observes the length of your spaceship to be	longer than / shorter than / the same as	his own spaceship.
Your friend observes the clock in your ship to be running	faster than / slower than / the same as	his own clock.
Your friend observes the mass of your ship to be	more massive than / less massive than / the same as	his own ship.

c) Why doesn't your relative motion affect the height and width of your spaceships?

S2.7) You and Greg own identical spaceships. Greg in his spaceship travels past you at high speed, while you are in your own spaceship. He tells you that based on his observations his ship is 20 m long and that the identical ship you are sitting in is only 5 m long. According to your observations [e.g. from your point of view]:

a) How long is your ship?

b) How long is Greg's ship?

c) What is the speed of your friend's ship?

d) While 60 minutes of time passes on your clock, how much time do you see pass on Greg's moving clock?

e) Does **time dilation** work just on mechanical clocks? Or is time dilation more fundamental – affecting every process associated with time?

S2.8) With the advent of an amazing new fuel source, an astronaut can travel at 0.866c. The first star that Dan decides to visit is Alpha Centauri – which Dan knows is 4.2 light years from Earth.

a) Dan needs to know how long it will take him to travel to Alpha Centauri (so he will know how much food and water to pack). Based on simple Newtonian physics, how much time should Dan expect to spend in his spaceship? [Hint: 1 light year is equivalent to 1 year × c. Complete your calculation using the variable c and don't bother plugging in the m/s value for c. Recall that distance = rate × time.]

b) Dan, however, realizes that because he will be travelling very, very fast the distance to Alpha Centauri will appear (from his point of view) to be **length contracted**.
  i. How far will Dan actually need to travel in his spaceship?

  ii. How much time (distance = rate × time) will this trip actually take Dan? (Dan's clock records this travel time.)

c) Observers on Earth watch Dan travel to Alpha Centauri. These observers see Dan travel the full 4.2 light years. Earth clocks record that the trip required the amount of time calculated in 'part a'. However, the Earth observers notice that Dan's clock recorded a very different amount of time travelled. As a result of **time dilation**, how much time do observers see pass on Dan's clock?

d) Do Dan and the Earth-bound observers agree on how much time passed <u>for Dan</u> during the trip?

e) If Dan left a twin sister behind on Earth, would Dan and his sister be the same age or different ages by the time Dan reached Alpha Centauri? If they are different ages, who is older?

Chapter S3

S3.1) Consider the spacetime diagram to the right.

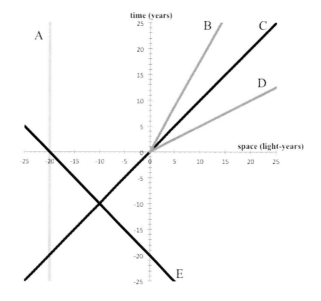

a) Which worldline (or worldlines) represents an object at rest? _____

b) Which worldline (or worldlines) represents light? _____

c) Which worldline (or worldlines) represents an object moving slower than light? _____

d) Which worldline (or worldlines) represents an object moving faster than light? _____
Why is this worldline not physically allowed for any object?

e) On the spacetime diagram above, draw the worldline of an object that is moving at the same velocity as B but in the opposite direction. Label this line "opposite direction."

f) On the spacetime diagram above, in the upper-right quadrant, draw the worldline of an object that is moving faster than B but slower than the speed of light. Label this line "faster object."

g) On the spacetime diagram above, in the upper-right quadrant, draw the worldline of an object that is moving slower than B but not standing still. Label this line "slower object."

h) On the spacetime diagram above, in the upper-left quadrant, draw the worldline of an object that starts at zero velocity but then accelerates to almost the speed of light. Label this line "accelerating."

i) On the spacetime diagram above, in the lower-right quadrant, draw the worldline of an object that starts near the speed of light but decelerates to zero velocity. Label this line "decelerating."

j) What does it mean when two worldlines intersect?

S3.2) The possible geometries for spacetime are **flat**, **spherical** and **saddle-shaped**. What are the characteristics of each type of geometry? *(Complete the chart given below.)*

	The sum of angles in a triangle	Behavior of parallel lines	Circumference of a circle	Straightest possible path
Flat				
Spherical				
Saddle-shaped				

S3.3) **A stretched latex sheet provides a 2-dimensional representation of spacetime.** Several tables will work together as one large group to experiment with a "spacetime simulator" provided by the instructor. Before starting, choose a group leader who will manage the group's experimentation. This leader should make sure the group follows the procedure below and discusses each question before moving on to the next step. The leader may want to assign roles to other members of the group such as simulator holders, ball rollers, Ping-Pong ball catchers, and scribe (to record group answers). *Be sure to hold the simulator level at all times.*

- a) Roll a Ping-Pong ball across the sheet several times, via different trajectories each time.
    - i. Describe the path(s) of the Ping-Pong ball across the sheet.

    - ii. What might the Ping-Pong ball represent?

- b) With the sheet held flat, place a weight in the center of the sheet. Roll one of the Ping-Pong balls across the sheet several times, via different trajectories each time.
    - i. Describe the path(s) of the Ping-Pong ball across the sheet.

    - ii. What might the weight represent?

    - iii. Can you place the Ping-Pong ball in orbit around the weight (at least for a short time)? If so, how?

c) This model is meant to simulate how mass warps spacetime in our universe.
   i. When placed on the latex sheet, the weights *curve* the sheeting. This is analogous to a mass curving spacetime. Based on what you observed in your model, what objects in the universe might cause curvature in spacetime? Which of those objects might cause the *most* significant curvature of spacetime?

   ii. How is **gravity** related to the curvature of spacetime?

   iii. What are the limitations to this model, when compared with the actual behavior of spacetime?

   iv. Why won't the Ping-Pong ball remain in orbit around the weight for very long? Specifically, what force (which is very common here on Earth but fairly rare in space) causes the Ping-Pong ball to lose kinetic energy and eventually crash into the weight?

S3.4) Imagine that you and a friend fly your spaceship to a nearby stellar-mass black hole. You put on a spacesuit and jump out of the spaceship, heading toward the black hole (as sketched below, *not to scale*). You carry with you a clock and a laser that emits yellow light. Your friend in the spaceship also has a clock and a yellow-light laser. You both shine your lasers toward the other.

a) From <u>your</u> point of view, what do you notice about your friend's clock? What color might your friend's laser look to you?

b) From your <u>friend's</u> point of view, what does your friend notice about your clock? What color might your laser look to your friend?

c) If you actually fell into the **event horizon** of the black hole (ignoring for now the tidal forces that would rip you apart), would your own clock and laser appear any different to you?

d) Could your friend see you fall into the event horizon of the black hole? *Explain why or why not.*

S3.5) Light will always follow a straight path – but sometimes that straight path is through curved spacetime! This results in the effect of **gravitational lensing**. When the lensing object is very massive, like a galaxy or a galaxy cluster, gravitational lensing can produce multiple images, arcs, or even Einstein rings from background light sources. In contrast, less massive lensing objects will result in **microlensing** events.

a) Why does a background light source appear brighter when a small, dim object (the gravitational lens) passes between the light source and the observer?

b) Sketch the approximate light curve (showing the change in brightness of the star over time) of the star shown below. A black hole is passing in front of the star from the observer's point of view, causing a microlensing event. Label the times A through E on your graph.

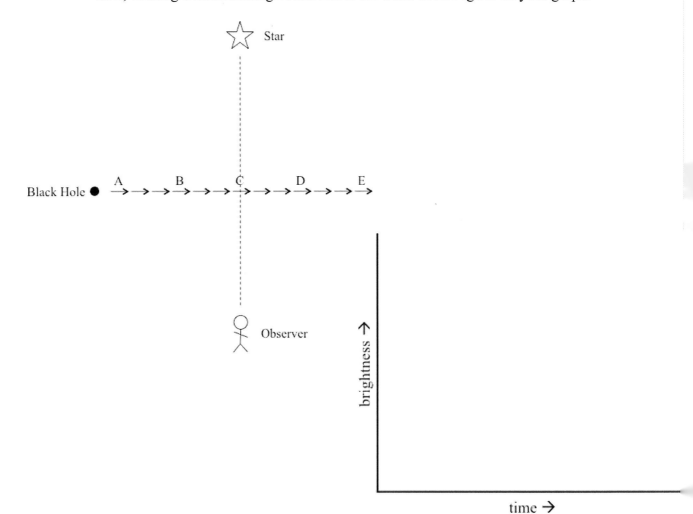

S3.6) On September 14, 2015, the Laser Interferometer Gravitational-Wave Observatory (LIGO) made the first direct detection of gravitational waves, which were produced when two black holes – which had been orbiting each other in a binary system – collided!

a) Why does the emission of gravitational waves cause the orbit of a binary system to decay (e.g. for the distance between the two objects to decrease with time)? After all, orbits are supposed to remain stable thanks to the conservation of energy and momentum!

b) Why doesn't the Earth-Sun orbital system decay due to the emission of gravitational waves?

c) After LIGO detects a gravitational wave signal, astronomers use telescopes to observe the region of the sky from which the gravitational wave signal was detected to try to pin-point the exact location of the event (since LIGO cannot provide an exact localization due to the nature of the detector). What might an observer see when two black holes collide? What might an observer see when two neutron stars collide?

Chapter 18

18.1) Compare the different kinds of compact stellar objects: **white dwarfs**, **neutron stars**, and **black holes**.

	White dwarf	Neutron star	Black hole
Produced by what type of star?			
Mass range			
Typical Radius			For a 1 M_{Sun} BH: For a 10 M_{Sun} BH:
Composition			
Supported by what type of pressure?			

18.2) White dwarfs!
   a) Over 5 billion years from now, our Sun's core will become a white dwarf. Will this white dwarf ever go novae or supernovae? *Explain your reasoning.*

   b) Compare **dwarf novae**, **novae** and white dwarf **supernovae**.

	Luminosity	Radiation produced by…	Can repeat? (yes or no)
Dwarf Nova			
Nova			
White Dwarf Supernova			

   c) In the 1920s, astronomers were engaged in a great debate about the size of the universe. Some astronomers believed the Milky Way was the full extent of the universe, while others believed that the Milky Way was just one "island" of stars in a much larger universe. The intrinsic luminosity of novae was part of this debate. If all novae had similar luminosities, then they must all be occurring within the Milky Way. Some astronomers, however, argued that there were actually two categories of novae: novae (~10^5 L_{Sun}) and supernovae (~10^{10} L_{Sun}). How much further away would supernovae have to be in order to appear to have the same **brightness** as novae?

18.3) Neutron stars!
   a) Draw a **pulsar**. Include and label: the neutron star, the axis of rotation, the magnetic field, and the beamed radiation.

   b) In science fiction, pulsars are sometimes used as starship navigation beacons (kind of like a galactic GPS system). The idea is that each pulsar has a distinct pulse period, and so by detecting 3 (or more) pulsars with known locations simultaneously a starship could triangulate its current galactic position. Based on the properties of pulsars, what might be the problems/limitations of this type of navigation system – which the system designers would need to keep in mind? (Explain at least 2 problems/limitations.)

   c) Compare and contrast **X-ray bursts** with **white dwarf novae**.

18.4) Black holes!
  a) Below is a schematic diagram of a black hole. Label the **event horizon**, **Schwarzschild radius**, and **singularity**.

  b) Out of pure curiosity, you decide you'd like to know what it feels like to cross the event horizon of a black hole. What type of black hole should you jump into in order to survive the trip through the event horizon – a "stellar mass" black hole or a "supermassive" black hole? *Explain the reason for your choice.*

  c) If you did survive the trip through the event horizon of a black hole…
     i. Would you be able to tell anyone outside the black hole about what you experienced? *Explain.*

     ii. What is your fate? In other words, what happens to you <u>after</u> you pass through the event horizon?

d) Why doesn't material from a companion star just fall directly into a black hole (or onto a WD or NS, for that matter)? Why is the accretion disk necessary?

e) Compare the accretion disks that form around protostars with those that form around compact objects (white dwarfs, neutron stars and black holes). How are they the same and how are they different?

f) Material from a companion star can fall onto the surface of a white dwarf and cause a novae, or onto the surface of a NS and cause an X-ray burst. Can a similar event occur in a black hole binary system? *Explain why or why not.*

g) A common misperception is that black holes are like vacuum cleaners, sucking up everything in sight. Why is this perception of black holes *incorrect*?

18.5) Consider the equation for the Schwarzschild radius of a black hole:

$$R_S = \frac{2GM}{c^2} = 3 \text{ km} \times \left(\frac{M}{M_{Sun}}\right)$$

a) Describe in words how the Schwarzschild radius of a black hole depends on the mass of the black hole.

b) If a typical human (~50 kg) could be crashed down enough to become a black hole, what would its Schwarzschild radius be?

c) What is the Schwarzschild radius of a 1 M_{Sun} black hole (in km)? _____ km

d) What is the Schwarzschild radius of a 10 M_{Sun} black hole (in km)? _____ km

e) What is the Schwarzschild radius of a 10^6 M_{Sun} black hole (in km)? _____ km

f) Once a black hole has formed, will it last forever if it has no source of additional mass (e.g. it's a lone black hole without a companion)? If not, explain what happens to this black hole over time?

18.6) The instructor will play video clips from science fiction TV shows or movies. In the space below, record what you believe to be scientifically accurate (fact) and inaccurate (fiction) about the portrayal of black holes in these clips.

Name of movie/TV show:

Science Fact	Science Fiction

Name of movie/TV show:

Science Fact	Science Fiction

Chapter 19

19.1) Draw a sketch of the Milky Way Galaxy. Include and label: the **bulge**, the **disk**, the **halo**, **spiral arms**, **globular clusters** and the approximate location of the solar system.

19.2) The schematic below shows the Milky Way Galaxy seen edge-on. Sketch in the orbit of several stars in each of the following locations: the disk, the bulge and the halo.

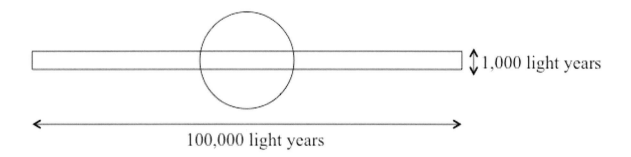

19.3) In the diagrams below, each large circle is uniformly filled with mass (each large circle is separate and doesn't affect the others). In each case, shade in the region of that mass that exerts a **net gravitational force** on the smiley face.

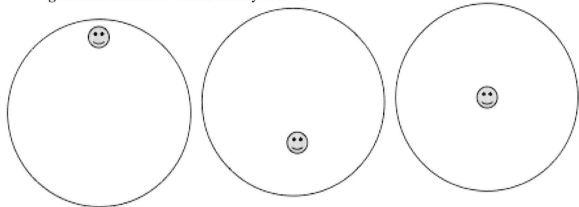

19.4) The solar system orbits the center of the Milky Way Galaxy at a speed of approximately 220 km/s. Gas on the outskirts of the Milky Way is observed to orbit the center of the Milky Way at approximately 250 km/s. Given what you know about Keplerian orbits (like in our solar system), why is this finding odd?

19.5) The mass of the Milky Way:

a) Use the orbital velocity law: $M_R = \dfrac{rv^2}{G}$, to calculate how much mass is in the Milky Way based on the orbit of gas at the outskirts of the Galaxy (radius ~ 50,000 ly). The gas orbits at a velocity of ~250 km/s.

b) The textbook calculates that the amount of mass in the Milky Way *within the Sun's orbit* is 10^{11} M_{Sun}. Based on your calculation in 'part a', how much mass must exist in the Milky Way in the <u>donut-shaped</u> region between the Sun's orbit and the orbit of the gas at the outskirts of the Galaxy?

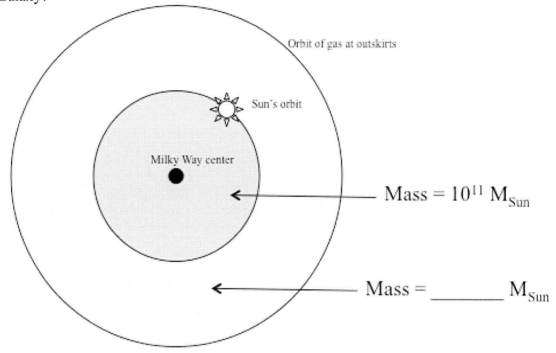

c) Look very carefully at either an artist's conception of the Milky Way or an image of the Andromeda galaxy. Do you find it surprising that there is an equal amount of mass in the inner region of the galaxy and the outer region of the galaxy?

19.6) Back in Chapter 1 (#1.2), you completed a calculation that indicated that galaxies are relatively large compared to the space between galaxies, while stars are exceedingly tiny compared to the space between stars within a galaxy. Individual stars, therefore, almost never collide with one another. Individual galaxies, however, collide quite frequently (especially within galaxy clusters).
   a) Are individual gas clouds within a galaxy large enough (compared with the space between them) to collide with one another on a regular basis?

   b) What happens when multiple gas clouds collide?

   c) Where are these types of collisions *currently* occurring within the Milky Way?

   d) Why do astronomers believe that more than one protogalactic cloud collided to form the Milky Way?

19.7) **Think like a scientist.** Your instructor will provide you with the number of a figure in your textbook – examine this figure carefully. This is a "time-lapsed" infrared image that shows the orbital pathways of several stars around the very center of the Milky Way galaxy. Each dot represents the exact position of the star during one yearly observation of the Galactic center – and 22 total observations were taken between 1995 and 2017. From this image, you can estimate the mass of the object at the center of the Milky Way.

a) From the image, **estimate** the orbital period (in years – you may need to extrapolate into the future!) and semi-major axis (in AU, using the scale bar on the image and a ruler) associated with each of the stars listed below. Using this period and semi-major axis, calculate the value of the central mass. Find the average of all 4 measurements.

To complete the calculations, you can use a form of Newton's version of Kepler's Third Law that has been manipulated to work with the units of years, AU and M_{Sun}.

$$M = \frac{a^3}{P^2}$$

M in solar masses (M_{Sun})
a in **AU**
P in **years**

Star	Period (yrs)	Semi-major axis (AU)	Central Mass (M_{Sun})
Light Blue			
Orange			
Red (circular)			
Dark Blue			
		Average:	

b) *Should* the 4 individual values you calculate for the mass at the center of the Milky Way be consistent with one another? *Explain why or why not.*

c) *Are* the 4 individual values you calculate for the mass as the center of the Milky Way consistent with one another? If the values are not consistent, why might the values differ? List and explain 2 possible reasons why the values may not be consistent.

   #1.

   #2.

d) Assuming that all the mass at the center of the Milky Way is contained within a single black hole, what is the Schwarzschild radius of this black hole? For this problem and all future problems, use the accepted value of 4×10^6 M_{Sun} for the mass at the center of the Milky Way.

e) Careful analysis shows that the central mass is confined to a radius of less than 50 AU. Could this mass just be made up of 4 million individual Sun-like stars (instead of the much more exotic sounding supermassive black hole)?

　　i. Calculate the total volume (in km³) in a sphere with a radius of 50 AU.

　　ii. Tightly packed spheres fill about 75% of a given space (the remaining space between the spheres is empty), so calculate how much of the volume in 'part i' could actually be filled with spherical objects.

　　iii. Calculate the total volume (in km³) of a typical Sun-like star.

　　iv. Based on 'part ii' and 'part iii', could 4 million Sun-like stars physically fit into a sphere with a radius of 50 AU?

v. Even if 'part iv' is mathematically possible, why is it unlikely that the central mass of the Milky Way is actually composed of 4 million individual Suns? Consider what might happen if you crammed that many Suns into an area smaller than our solar system…

Chapter 20

20.1) Compare **spiral**, **elliptical** and **irregular** galaxies.

	Spiral	Elliptical	Irregular
General description of shape			
Color			
Contain young, massive stars?			
Contain old, low-mass stars?			
Contain cool gas and dust?			
Ongoing star formation?			

20.2) The light curves (apparent brightness – in magnitude units – versus time) of four Cepheid variable stars are given below[5].

a) What is the pulsation period of each Cepheid?

Cepheid A _____

Cepheid B _____

Cepheid C _____

b) Rank the Cepheids in order of increasing <u>intrinsic</u> luminosity.

Lowest luminosity ____ ____ ____ Highest luminosity
Explain your reasoning:

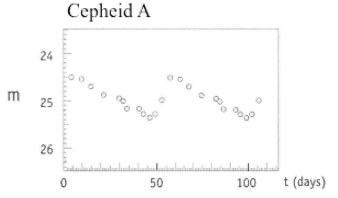

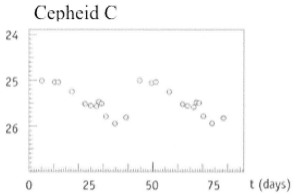

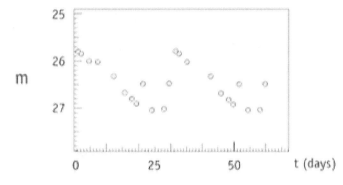

---

[5] Adapted from Bacher & Christensen "The Distance to Messier 100 as Determined by Cepheid Variable Stars" (http://www.astrosociety.org/edu/publications/tnl/57/realastro3.html)

20.3) The **distance ladder** (or **distance chain**) builds up step-by-step to allow astronomers to measure the distance to objects throughout the universe. How would the rungs up the ladder be affected if astronomers made a mistake at a lower rung?

   a) Imagine that astronomers have made a mistake in the Parallax measurements to the Hyades. They have accidently measured the Hyades to be much <u>further</u> than it actually is! Would astronomers believe that the stars in the Hyades were more luminous or less luminous than in reality? *Explain your reasoning.*

   b) Since the astronomers have misidentified the intrinsic luminosity of the stars in the Hyades, they've also misidentified the intrinsic luminosity of the Cepheids in the Hyades. If they attempt to use the period-luminosity relationship of Cepheid variable stars in the Hyades to measure the distance to other galaxies, will the measurements underestimate or overestimate the distance to those galaxies? *Explain your reasoning.*

20.4) What is the difference between the Doppler shift and cosmological redshift (since both shift the observed wavelength of spectral lines)? Specifically, what causes each effect?

20.5) Bob says to you, "Hubble's Law ($v = H_0 \times d$) seems to be suggesting that the Milky Way galaxy is in very special – we must be at the center of the universe! How else could it appear that almost all galaxies move away from us, with the more distant galaxies moving away at a faster rate?" Do you agree or disagree with Bob's statement? If you disagree, how would you correct Bob?

20.6) Kristin asks you, "If the universe is expanding, it would seem that the spaces between all objects must be getting bigger. So wouldn't our solar system or galaxy eventually be pulled apart? And what about the space between the atoms or even the subatomic particles in my body? Is my body being stretched apart, too?" How would you respond to Kristin's questions?

20.7) The Hubble Constant is measured to be approximately 22 km/s/Mly.
   a) From this value, calculate the age of the universe.

   b) If the Hubble Constant was actually 30 km/s/Mly, would the universe be younger or older than what you calculated in 'part a'? *Explain your reasoning.*

   c) Imagine that two astronomers are arguing over the Hubble Constant. Jennifer believes that the Hubble Constant must be 15 km/s/Mly, while Fred believes that the Hubble Constant must be 10 km/s/Mly. Who believes that the universe is older? *Explain your reasoning.*

20.8) **Think like a scientists.** Launch the Appreciating Hubble at Hyper-speed (AHaH) program[6]. This software combines galaxy images from the Hubble Ultra Deep Field with spectroscopically measured redshifts of the galaxies to create an accurate three-dimensional interactive simulation of a slice of the universe. Try zooming into and out of the image, and panning from side to side.

a) *In each redshift range* listed in the table below, identify 10 galaxies and classify each as spiral, elliptical, or irregular. Record the Object ID of each galaxy in the table (to avoid double-counting any galaxies).

Redshift range	Spiral	Elliptical	Irregular
0.0 – 0.3			
0.3 – 0.5			
0.5 – 1.0			
1.0 – 1.5			
1.5 – 2.0			
2.0 – 2.5			
2.5 – 3.0			
3.0 – 4.0			

---

[6] *Adapted from Mechtley, Will, & Windhorst "Appreciating Hubble at Hyper-speed Tool Exercises"* (http://www.asu.edu/clas/hst/www/ahah/exercises.html)

b) Make a histogram of your data.

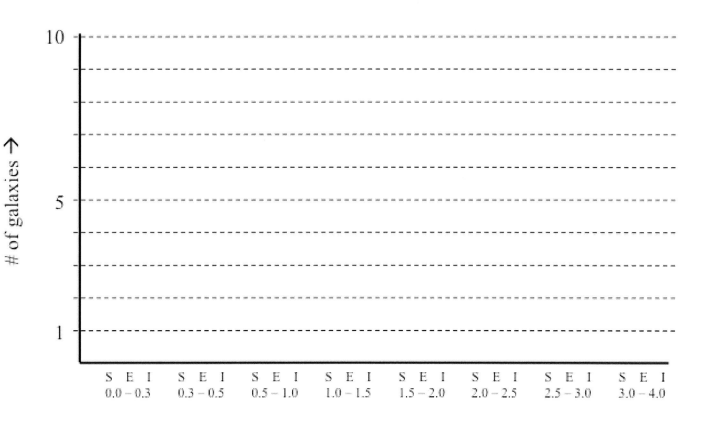

c) Based on your data:

Which type of galaxy is most common in the early universe? _____

Which type of galaxy is most common in the universe today? _____

d) Current models of galaxy formation start with small irregular blobs of gas and young stars merging and forming larger galaxies. What <u>evidence</u> did you see in this experiment to support or refute this idea?

Chapter 22

22.1) Think back to the three hallmarks of science. Does the **Big Bang** theory fulfill each hallmark? *Explain your reasoning*.

    Hallmark #1:

    Hallmark #2:

    Hallmark #3:

22.2) The early universe underwent incredible changes in astoundingly short periods of time. What is the fundamental reason that these changes occurred?

22.3) What three key features of the universe can be explained by inflation? Describe each in your own words.

    #1.

    #2.

    #3.

22.4) Daniel says, "Inflation can't have happened! For the universe to increase in size by a factor of 10^{30} in just 10^{-36} seconds would mean that matter in the universe would have to move at faster than the speed of light. Relativity clearly states that this is impossible – no matter or information can travel that fast!" Do you agree or disagree with Daniel's statement? If you disagree, how would you correct Daniel?

22.5) **Cosmic microwave background!**
   a) Use the spectrum of the cosmic microwave background (your instructor will provide you with the number of the figure in your textbook), and Wein's Law to calculate the current temperature of the universe. Note that 1 mm = 10^6 nm.

   b) The stretch factor for light is $1 + z = \frac{\lambda_{observed}}{\lambda_{emitted}}$. If the light that makes up the CMB initially had a peak wavelength of 970 nm, by what factor has that light been stretched in the last ~ 14 billion years?

22.6) **Think like a scientist.** The **critical density** can be thought of as the density required for galaxies within the universe to be able to recede from the universe at exactly the escape velocity of the universe itself!

  a) If the universe was <u>denser</u> than the critical density, what might happen to galaxies in the universe?

  b) If the universe was <u>less dense</u> than the critical density, what might happen to galaxies in the universe?

  c) Complete the following steps to calculate *algebraically* the critical density of the universe.
  i. Using Hubble's Law, write down the recession velocity of a galaxy with distance d.

  $v_{recession} =$

  ii. Using the escape velocity equation, write down the escape velocity of a galaxy escaping the universe (which has mass M) from a distance of d.

  $v_{escape} =$

  iii. We cannot quote a value for the mass (M) of the universe. Instead, we will need to work with values that can be measured locally. Density is such a value. Recall that *density = mass/volume*. Algebraically, how much mass (M) is enclosed in a spherical area of radius d with a density of ρ_{crit}? Substitute this value into the equation for escape velocity from 'part ii'.

  $M =$

  $v_{escape} =$

iv. Since at exactly the critical density the recession velocity will be equal the escape velocity, set $v_{recession}$ (from 'part i') = v_{escape} (from 'part iii'). *Algebraically*, solve the equation for ρ_{crit}.

$\rho_{crit} =$

v. Consult with an instructor to make sure that your derivation of the critical density is correct. If so, calculate the numerical value for the critical density of the universe (in units of kg/m^3). For H_0 use the value of 22 km/s/Mly = 2.3×10^{-18} s^{-1}.

$\rho_{crit} =$ _____ kg/m^3

vi. The critical density is equivalent to approximately how many <u>hydrogen atoms</u> per cubic meters of space? Does the answer surprise you?

vii. All measurements completed by astronomers to this point indicate that the shape of the universe if **flat**. What, then, must the actual average density of the universe be? *Explain your reasoning*.

Chapter 23

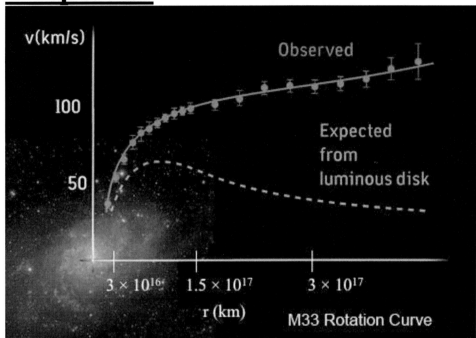

(Image Credit: NOAO, AURA, NSF, T.A.Rector.)

23.1) The observed (solid) and predicted (dashed) rotation curves of M33 are show above.
   a) Given M33's observed rotation curve, how much mass (in M_{Sun}) is there in the galaxy inside of 3.3 kly (3×10^{16} km), 16.5 kly (1.5×10^{17} km) and 33 kly (3×10^{17} km)? Be cautious with units in your calculation!

3.3 kly = _____ M_{Sun} 16.5 kly = _____ M_{Sun} 33 kly = _____ M_{Sun}

b) How did astronomers measure the rotation velocity of the galaxy out to 33 kly, when the image shown above (an optical light image) only has an extent of about 16.5 kly?

c) Astronomers have calculated the curve (labeled "expected from luminous disk") that would be expected if the galaxy were comprised only of visible material (e.g. stars and gas). If the galaxy was only comprised of visible material, how much mass (in M_{Sun}) would there be inside of 3.3, 16.5 and 33 kly?

3.3 kly = _____ M_{Sun} 16.5 kly = _____ M_{Sun} 33 kly = _____ M_{Sun}

d) At each radius (3.3, 16.5 and 33 kpc), which mass dominates, that of the visible material (stars and gas) or that of the unseen matter (**dark matter!**) in the galaxy?

23.2) Besides flat galaxy rotation curves, what other evidence points to the existence of a large amount of dark matter in our universe?

23.3) **Think like a scientist.** What exactly is dark matter? Astronomers considering this question developed several different hypotheses. One initial explanation for dark matter was that it might be made up of many small, dark objects composed of <u>normal</u> (baryonic) matter. These objects must be very cold so that they emit almost no thermal radiation. Possible examples include black holes, cold neutron stars, cold white dwarfs, brown dwarfs (failed stars), or even "rogue" (or "orphan") planets. These objects are collectively referred to as MACHOs (MAssive Compact Halo Objects).
   a) The mass of the Milky Way within the Sun's orbit is approximately 10^{11} M_{Sun} – most of which is dark matter. If dark matter was actually composed of MACHOs, how many MACHOs must there be in every cubic light year of space in this region of the Galaxy? Assume individual MACHOs had a typical mass of 0.05 M_{Sun}. Recall that the Sun is about 27,000 light years from the center of the Galaxy, and that dark matter is *not* confined to the Galactic plane but is instead spread out in a sphere.

b) Based on the number of MACHOs required per cubic light year, does it seem reasonable that MACHOs might be a viable dark matter candidate? *Explain your reasoning.*

c) In order to test whether or not dark matter is composed of MACHOs, astronomers have carried out surveys looking for microlensing events. What is observed during a microlensing event?

d) Why would searching for micolensing events be a good way to detect MACHOs in the Milky Way?

e) MACHOs have, in fact, been identified through microlensing surveys. However, only a handful of MACHOs were found in these surveys – far too few to account for all the dark matter assumed to be in the Milky Way. Based on our understanding of the Big Bang, why was this outcome not surprising?

23.4) As you watch the Dark Matter "NOVA scienceNOW" video, consider the following questions:
   a) What is Dr. Figueroa's Dark Matter detector made from?

   b) Why must the Dark Matter detector be cooled to 50 mK?

   c) Why is the Dark Matter detector buried half a mile underground in a mineshaft?

   d) How many particles of Dark Matter has Dr. Figueroa detected so far?

   e) How might Dark Matter have helped galaxies to form?

23.5) How did tiny random fluctuations in the early universe lead to the current large-scale structure of the universe – including the superclusters and voids?

23.6) **Think like a scientist.** Let us first imaging a universe composed of just <u>matter</u> (both regular and dark). In this universe, there is a "battle" between the expansion imparted by the Big Bang and the gravitational attraction of all the matter in the universe. This gravitational attraction, in a sense, is trying to draw the whole universe together. The fate of the universe, therefore, would be decided by who "wins" the tug-of-war: expansion or gravity. To quantify the relative density of the universe in comparison with the critical density, astronomers define the following parameter:

$$\Omega = \frac{\rho_{actual}}{\rho_{critical}} = \frac{actual\ density\ of\ the\ universe}{critical\ density\ of\ the\ universe}$$

a) In this matter-only universe, what would happen to the universe in the long term if the universe had a density that was <u>greater</u> than the critical density ($\Omega > 1$)? This is often referred to as a "recollapsing" (or "closed") universe.

b) In this matter-only universe, what would happen to the universe in the long term if the universe had a density that was <u>equal</u> than the critical density ($\Omega = 1$)? This is a "critical" universe.

c) In this matter-only universe, what would happen to the universe in the long term if the universe had a density that was <u>less</u> than the critical density ($\Omega < 1$)? This is often referred to as a "coasting" (or "open") universe.

d) What about a universe that is <u>empty</u> ($\Omega = 0$)? This universe would contain absolutely nothing, no stars, gas or galaxies. What would happen to this universe in the long term?

e) How would an empty universe's expansion differ from that of a coasting universe? Keep in mind that the coasting universe basically has just a little bit of matter, while the empty universe has none.

f) In the diagram below, label the recollapsing, critical and coasting universe. There is an additional type of universe in the diagram, which we will return to later in this exercise.

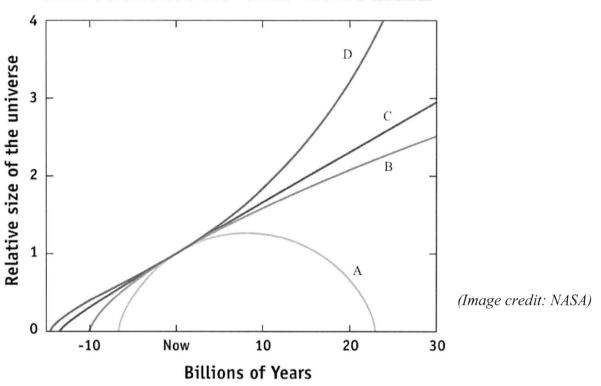

(Image credit: NASA)

g) Which universe would be the youngest? A recollapsing, critical or coasting universe?

h) If the circles below represent the *current* size of the universe, sketch how the size of the universe would change in the future in each universe (sketch 3 snapshots of each future universe).

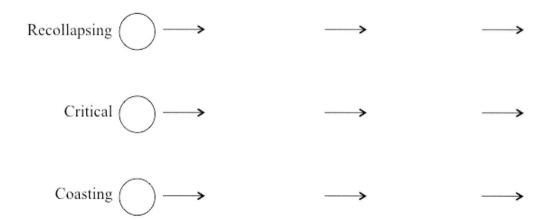

i) Measurements of the universe suggest that the geometry of the universe is flat (recall that inflation helped to explain this flatness). This suggests that the actual density of the universe must be equal to the critical density ($\Omega = 1$). But when astronomers measure the density of all the matter in the universe (again, both regular and dark matter) they come up very short! In fact, matter accounts for only about one quarter of the critical density. What do astronomers believe makes up the remaining ~75% of the density of the universe?

j) Consider curve D in the "Expansion of the Universe" plot on the previous page – label curve D the "accelerating" universe. Describe in detail how the size of that universe changes with time.

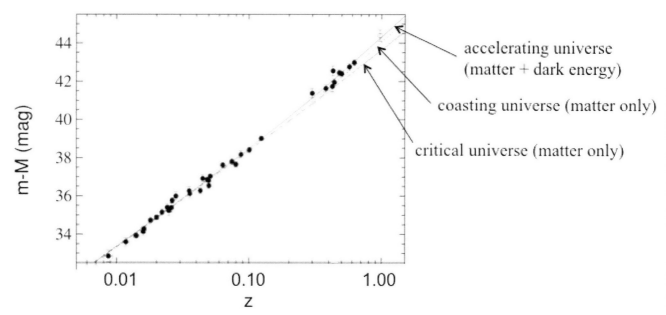

k) The figure above is from Riess et al.'s 1998 *Astronomical Journal* article, which was one of two papers that simultaneously announced the observational evidence for an accelerating universe. The data shows the distance modulus (in magnitude units) of supernovae from both the local (low z) universe and the more distance (larger z, further back in time) universe. These data points have been plotted over curves drawn for 3 different universes: accelerating, coasting, and a matter-only critical universe. The fact that the data points were most consistent with an accelerating universe was a groundbreaking discovery! Why were the observations of the *highest redshift supernovae* critical for making this discovery?

l) The discovery of the acceleration of the universe depended strongly on the measurement of the brightness of very, very distance supernovae. Also, this discovery requires the assumption that Type Ia supernovae are true standard candles (e.g. have the same intrinsic luminosity no matter their place in the universe). Based on what you know about the contents of the universe, and how stars change over many generations, what sources of error/uncertainty must have concerned astronomers in making these measurements and assumptions?

23.7) A word of caution! Do not confuse dark matter with dark energy. Also, do not confuse expansion with acceleration.
   a) What is **dark matter**?

   b) What is **dark energy**?

   c) Describe how the size of <u>our</u> universe changes over time.

   d) What is the ultimate fate of <u>our</u> universe?

23.8) Imagine that humans still live in the Milky Way galaxy when the universe is 100 billion years old (we must have colonized some other solar systems!). It is likely that the Milky Way would have collided/merged with several other galaxies by that point.

    a) What would the night sky look like to the naked eye? How would it be different from today and how would it be similar to today?

    b) What would have changed about the contents and processes occurring in the Milky Way galaxy?

    c) If all knowledge of cosmology had somehow been lost to the human race, could humans reacquire that knowledge when the universe is 100 billion years old? In other words, with the right equipment could humans detect other galaxies and reformulate the Big Bang theory? *Explain why or why not.*

A Publication from DLC, LLC

The Fundamentals of Counseling: *A Primer*

e-mail: **Primer@dlc.email**

© 1995
7th Edition – **August 2014**
ISBN: 978-0-692-25904-7
DLC, LLC